D1000402

The Year of the Dinosaur

The Year of the Dinosaur

Edwin H. Colbert
Illustrated by Margaret Colbert

Charles Scribner's Sons
NEW YORK

Library of Congress Cataloging in Publication Data

Colbert, Edwin Harris —
The year of the dinosaur.

Bibliography: p.
Includes index.
1. Brontosaurus. 2. Dinosauria. 3. Paleontology—Mesozoic.
I. Title.
QE862.D5C683 568'.19 77-12919
ISBN 0-684-15168-5

This book published simultaneously in the United States of America and in Canada—Copyright under the Berne Convention

1 3 5 7 9 11 13 15 17 19 V/C 20 18 16 14 12 10 8 6 4 2

Printed in the United States of America

CONTENTS

PREFACE vii

PROLOGUE xi

FIRST MONTH: The Hunted and the Hunter 1

SECOND MONTH: The Forest and the Swamp 13

THIRD MONTH: The Land of Giants 33

FOURTH MONTH: The Sun and the Shade 47

FIFTH MONTH: The Tread of the Giant 65

SIXTH MONTH: The Constant Wanderer 75

SEVENTH MONTH: The World Outside 85

EIGHTH MONTH: The Return 97

NINTH MONTH: The Nest 107

TENTH MONTH: Confrontations 119

ELEVENTH MONTH: The Flood 131

TWELFTH MONTH: The Hatchlings 141

NOTES 155

SELECTED AND ANNOTATED BIBLIOGRAPHY 163

INDEX 167

PREFACE

THIS BOOK IS the story of a year in the life of a large dinosaur (the central character), and of a group of similar dinosaurs with which the protagonist was associated. As such, it is a story concerned to a large degree with dinosaurian behavior. But how does one write about the behavior of animals that lived 140 million years ago, particularly since those animals were quite unlike any living things on the earth today? It would appear to be an impossible task.

It is not impossible, but it is not easy, either. It is in effect a problem of reconstruction—of restoring long-extinct animals as credible living creatures without taking a high-flown path into the realm of science fiction. This re-creation of what ancient animals did, in a world very different from ours, must be based upon various lines of evidence.

First, as far as this story is concerned, there is the evidence of the fossil bones. Bones do not reveal behavior, but they can be clues to behavior. For example, the great size of the dinosaur bones that represent the participants in this narrative are the fossil remains of ancient reptiles of immense bulk and weight. Their very size indicates that these animals must have behaved in certain ways, governed by the laws of physics, as are modern terrestrial giants such as elephants. The size of the fossil bones, and correlatively of the animals for which they were once the supports, says something about the physiology of the long-extinct dinosaurs, which in turn gives clues to some of their behavior patterns. In addition, it is possible that the internal structure of these bones may provide clues to dinosaurian physiology, particularly body temperatures. Furthermore, the manner in which the skeletons of the dinosaurs are put together—in other words, the articulations of bone with bone throughout an osteological framework of astounding size and remarkable complexity—can furnish evidence as to how the giant reptiles moved about and how they behaved under various circumstances. So in order to reconstruct a dinosaur as an active, moving animal one must devote detailed attention to anatomy.

Then there are the footprints, which often form long trackways. They are the imprints in stone of what happened at a particular moment, millions of years ago, providing a sense almost of contemporaneity with the animals that made them. Thus, the remarkable footprints from Texas, on which some events in this story are based, show very clearly just what several dinosaurs were doing on a certain day, when that part of Texas to the west and south of Dallas was a tropical land bordered by a tropical sea. Dinosaur footprints are fairly numerous, and they are valuable when properly interpreted.

Fossil dinosaur eggs have also been found. They are not especially numerous, but they do occur. And they give us information, at least in some instances, about how dinosaurs built their nests, and how they perpetuated their particular species.

Even the conditions under which the carcasses of dinosaurs were buried and subsequently fossilized can supply information about the probable habits of the long-extinct reptiles. Dinosaur skeletons in crossbedded sandstones deposited by rivers reveal that the ancient reptiles frequented riverbanks. The association of shells or fossil wood or pollen with such skeletons provides additional facts about the habits of the reptiles, especially concerning their choices of habitats. The information derived from some fossil finds may be even more precise. For example, two interlocked skeletons of dinosaurs were found in Mongolia, which seem to show that the reptiles perished while fighting.

Of course we can make extrapolations from the behavior of modern reptiles to interpret what some of the behavior of dinosaurs may have been like. For this purpose the studies of crocodilians, the closest living reptilian relatives of the dinosaurs, are most useful. So also are studies of large turtles. It is reasonable to suppose that in many instances dinosaurs may have behaved in much the same way as those large modern reptiles.

The comparison with modern animals need not be restricted to reptiles. There is good evidence that birds are direct descendants of certain dinosaurs; therefore some aspects of bird behavior, such as locomotion and feeding in ostriches, may reflect certain facets of dinosaurian behavior. In that respect, however, perhaps more useful comparisons are to be made with some of the giants among modern mammals, especially the elephants, as already mentioned. Some of the problems of living for dinosaurs must have been rather similar to those experienced by modern elephants.

It may be argued that the comparison of reptile with mammal is not justified. But perhaps it is. The dinosaurs were very special reptiles which may have had some of the attributes of "warm-blooded" mammals. (Whether the dinosaurs were "cold-blooded," or ectothermic, as are all modern reptiles, or independently "warm-blooded," or endothermic, is a debatable question to be considered later in this book; but they may very well have had various behavior patterns generally characteristic of large endotherms.)[1] Of course the dinosaurs had a primitive, reptilian brain, and for that reason much of their behavior must have been very different from that of large, intelligent mammals. Nonetheless, some similarities perhaps can be justified.

In summary, there are various lines of evidence that enable us to reconstruct many of the activities that typified the life-style of the dinosaurs. The present story has been written on the basis of such evidence, in an attempt to present a picture of what life was like for some of the greatest of the dinosaurian giants.

PROLOGUE

BRONTOSAURUS, the principal character in this story, and the other dinosaurs forming the herd to which she belonged were members of a single dinosaurian species, which was but one among hundreds of species of dinosaurs now known. Therefore, to speak of the year of the dinosaur is to speak in general terms. Although it might be more accurate to entitle this book *The Year of the Brontosaur*, no particular advantage would be gained by such specificity, especially in view of the fact that the narrative includes various dinosaurs other than Brontosaurus.

The dinosaurs were reptiles, now extinct, that lived during the Mesozoic Era of earth history, a time span that began about 225 million years ago and ended some 65 million years ago. The dinosaurs arose during the Triassic Period, with a duration of some 35 million years, the first of the three geologic periods comprising the Mesozoic Era;

they evolved and multiplied during the Jurassic Period, which began about 190 million years ago and lasted approximately 55 million years, the middle division of the Mesozoic; and they reached the climax of their long and complex history during the Cretaceous Period, the final and seemingly the most extended Mesozoic period, with a duration of about 70 million years.[2] With the end of the Cretaceous the dinosaurs became extinct.

The dinosaurs are contained within two great orders of reptiles, the Saurischia and the Ornithischia, which followed separate lines of evolutionary development. Within those two orders, some eight suborders (the number varies slightly according to the classifier) express the divergence among those reptiles. Contained within the Saurischia are the theropods, the one group of carnivorous dinosaurs; the prosauropods, primitive ancestors of the giant sauropods; the sauropods, largest of the dinosaurs. Among the Ornithischia are the ornithopods, characterized especially by the Cretaceous duck-billed dinosaurs; the stegosaurs, the plated dinosaurs; the ankylosaurs, the armored dinosaurs; the pachycephalosaurs, the dome-headed dinosaurs; the ceratopsians, the horned dinosaurs. Even the names suggest the variety that marked dinosaurian evolution through a span of more than a hundred million years. That great array of dinosaurs dominated the continents of the Mesozoic world.

From the wide roster of dinosaurs certain sauropods were chosen for this story. Moreover, the choice was narrowed to sauropods of the genus *Apatosaurus*, commonly known as brontosaurs. *Apatosaurus* lived in western North America (at least that is where the fossils have been found, in the Morrison formation) during the late Jurassic Period. So those facts have determined the setting for the story.

Some of the dinosaurs contemporaneous with *Apatosaurus* in North America occur elsewhere: for instance, the gigantic sauropod *Brachiosaurus* is found in Tanzania, as well as in North America. So perhaps *Apatosaurus* ranged beyond the limits of present-day North America, when the continents were much closer to each other than they are now, and large dinosaurs could roam freely far and wide. However, this is a story of western North America.

Although it is essentially a story of the late Jurassic Period, certain liberties with time have been taken. The gigantic footprints that form a basis for the beginning of the narrative occur in Lower Cretaceous beds in Texas. Nevertheless they are essentially the kinds of footprints that would have been made by *Apatosaurus* (Brontosaurus) and by *Allosaurus*. They were made in the same type of environment that might have been inhabited by Brontosaurus and contemporaneous dinosaurs. Therefore it is quite legitimate to use them as a basis for a part of this account.

The northern sea described in the story is the Sundance Sea, which may have existed somewhat before the time when Morrison dinosaurs lived, although there is reason to think that the Sundance formation may be in part contemporaneous with the lower part of the Morrison formation. Again, it is felt that the inclusion of the sea in this story is justified.

The story is a general account of a general time—a time encompassing the latter part of Jurassic history and the beginning of Cretaceous history, when sauropod giants ruled the land as they had not ruled it before (because before this time they had not become such giants) or after (because after this time the marvelous parade of varied Cretaceous dinosaurs encroached upon the world of the sauropods, to diminish their dominance). It tells of a time when extreme giantism, as exemplified by the great sauropods, had reached its apogee, before the smaller but more varied dinosaurs of later Cretaceous history established their ascendancy across the land. This is a story of a time when there were giants on the earth, the like of which were never seen again. Finally, it is the story of one year in the life of some of those giants.

The Year of the Dinosaur

FIRST MONTH

The Hunted and the Hunter

THE WARM SUN SHONE brightly on the low seashore, accentuating the dark green jungle along the beach and creating a million bright points of light at the tips of the little waves that lapped across the shallows. The beach was white, and the sea floor, plainly visible at close range beneath a bare three or four feet of crystal-clear water, was white, too. It was a vast expanse of soft mud that made an almost level bottom from the water's edge far out into the embayment. The reflected light from the beach and from the surface of the water covering the shoals was dazzling, and if there were eyes to see, they might have found relief in turning from the water and the beach toward the jungle that stretched mile after mile along the shore.

It was a thick and a rather monotonous forest, an ancient forest, that extended back from the beach. Towering into the sky were straight-trunked conifers, some with radiating branches from base to tip, very much like many of the pine trees with which we are familiar, and some with terminal crowns, like huge, green, uneven bouquets, carried on the tips of bare, columnar trunks. Beneath these trees, and often forming their own lesser jungles, were the tree-ferns, standing ten to thirty feet in height. They were what the name implies: ferns in the form of trees, ferns branching out from the tops of scaly trunks, with their greenish yellow fiddleheads carried high above the ground. Such tree-ferns are to be seen today in tropical and semi-tropical jungles around the world.

Even shorter than the tree-ferns were the abundant cycads, with big, globular boles, from the tops of which radiated long palmlike fronds. On the boles of the cycads were bright rosettelike fructifications, or "flowers," affording flashes of color in the green landscape. The cycads were plentiful, and their flowers formed a broad, interrupted understory of color beneath the green vegetation. There were other cycads, too; the Williamsonias, looking like stubby trees with scaly trunks and crowns of upstanding fronds.

In wet places, in little marshes and sloughs, were giant horsetails or scouring rushes, their jointed, cylindrical stems decorated at regular intervals by whorls of thin branches protruding from the joints. Those primitive plants, thrusting several feet or more vertically from the ground, the diminutive relatives of which can still be found in wet environments, added their green color to the scene, except that their tips were russet, thereby supplementing in a very modest way the variety of color provided by the cycad flowers. On the ground were various low, spreading plants, among them numerous ferns.

It was a landscape that would seem strange to our eyes, for it existed more than a hundred million years ago.

Brontosaurus,[3] one of the largest of the dinosaurs, crashed through that forest toward the beach. What led her from the shade of the jungle into the intense light of the beach and the shallow water is hard to say. Perhaps she was taking a shortcut across an embayment to the jungle on the other side. At any rate, she plodded in a purposeful manner directly toward the water.

Her huge form loomed large against the backdrop of greenery from which she had just emerged. She was more than seventy feet long, much of her length consisting of a long, rather sinuous neck and an even longer muscular tail. Her hips, the highest part of her body, were fifteen feet above the ground—as high as the crowns of the tree-ferns—raised by her long, thick hindlimbs, elephantlike legs capable of supporting the larger fraction of her thirty-ton body.[4] In contrast, her forelimbs were relatively short, even though they, too, were heavy and strong. Consequently her shoulders were lower than her hips, so that the line of the back sloped in a great sweeping curve forward from its high point down to the shoulders and on to the base of the neck, and backward down to the base of the tail and to its extremity. Her tough, leathery skin was creased with many folds and wrinkles.

Both the forelimbs and the hindlimbs terminated in broad, somewhat rounded, padded feet, like the feet of an elephant, but different, too. The forefeet were rounded in front; there was a large inner claw on each foot, while the front border may have been decorated with blunt nails. Of the latter feature we cannot be sure. But the hindfeet, much larger than the forefeet—each a yard (or a meter) long, and elliptical—were armed with three huge claws, borne on the first, second, and third toes. Those claws, which slanted outward, dug into the earth as Brontosaurus walked along with ponderous strides.

As she walked she swung her head from side to side in long horizontal arcs, thus scanning the scene from differing points of view. To say that she swung her head widely is no exaggeration, for her neck was at least fifteen feet in length from the shoulders to the base of the skull. It was a strong and very useful neck.

At the end of the boomlike neck was the head: a fairly large, bony structure in human terms, but a remarkably small head in relation to the bulk of her body. From the sides of the head her large eyes stared to the right and the left and forward, comprehending the scene in reptilian terms that will be forever unknown to us. Directly in front of the eyes and slightly higher in elevation were the large nostrils, which flared open in the form of round, muscle-rimmed apertures, or closed into slits. Those slits would keep out the water when the head was lowered beneath its surface. Finally, making a horizontal line from the front to the back of the head, were the jaws, in the forward portions of which were set a limited number of rather uneven, peglike teeth, all alike. They appear to have been teeth for cropping vegetation, perhaps leaves of trees and shrubs, perhaps underwater plants, dredged up from shallow rivers, lakes, and marshes.

As she walked along, her heavy tail dragged on the ground, leaving a groovelike trail behind her.

She left the forest, crossed the beach, and entered the shallow water in long, splashing strides. With each step her huge feet sank deep into the muddy bottom, leaving impressions as big as washtubs. As each foot pressed down, bearing the great weight which it partially supported, the mud welled up around it, to form a raised rim bordering the depression that was left when the foot was lifted. Within the rims at the front of the hindfoot impressions were deep marks made by the claws, three for each of these tracks.

Thus she waded through water that was too shallow to reach her belly, so that her full weight pressed down on her feet to leave telltale footprints marking her course, yet deep enough so that her tail floated. Consequently there was no groovelike mark of the tail drag between the footprints, as there might have been if her journey had been recorded on dry land.

Her gait was unhurried, for there was no reason for her to try to make haste. Each foot was firmly planted into the muddy bottom; each foot was lifted out of the ooze with a sucking sound. She waded in a straight course through the shallow water, making for the somewhat distant shore which she could see on the other side of the embayment.

As Brontosaurus splashed through the shallow water of the embayment, Allosaurus[5] came out of the forest that fringed the beach. He was a very different dinosaur from Brontosaurus—not so large, but decidedly menacing in appearance. The newcomer on the scene was a carnosaur, one of the giant carnivores that preyed upon other large reptiles, including any dinosaurs that they might be able to overcome. The carnosaur stood high on his hindlimbs, which were remarkably birdlike in form. Indeed, the long hindlimbs of Allosaurus, although massive and muscular in their upper portion, were comparatively slender below the ankles—as slender as they might be for the support of a body weighing some three tons—and they terminated in sharp-clawed, three-toed feet, closely resembling in form the feet of a large bird. The feet of Allosaurus might well have resembled the feet of large birds, for they functioned in almost exactly the same manner as the feet of some large birds with which we are familiar. This dinosaur could cross the ground in long strides; in fact he could run with considerable speed.

The forelimbs of Allosaurus were relatively small and quite useless for locomotion. They were, in effect, arms and hands, useful for grasping and holding things, for restraining any unfortunate victim the carnivore might capture. Each hand was furnished with three large, sharp, curved claws, terminating in the thumb and two fingers. The inner claw on the thumb was the largest; the other two were progressively smaller. Together they made a set of fearsome hooks of undoubted efficiency.

Since Allosaurus was a strictly bipedal reptile, his body was pivoted at the hips; and since he did not stand in an upright position, as a man does, he had a long, heavy tail that served among other things to counterbalance the weight of the body. We see something of the same arrangement today in a kangaroo, although the kangaroo is a hopping animal rather than a runner, as was this dinosaur.

A most striking feature of the gigantic carnivore was his huge head, carried some fifteen feet above the ground and set upon a comparatively short, strong neck. The head was so large that it might have appeared to be almost out of proportion to the rest of the animal, but there was a reason for its great size. It provided Allosaurus with a formidable set of strong jaws, which were armed with a dazzling array of sharp, saberlike teeth; the thin edges of each tooth were serrated for cutting its food. The jaws and teeth demonstrate a pattern for predation on a large scale.

Allosaurus—like Brontosaurus, which had preceded him out of the forest—was clothed with heavy, tough, leathery skin.

So here they were, the hunted and the hunter, playing out an ancient drama on a gigantic scale, for Allosaurus had seen Brontosaurus wading across the shoals and had decided to pursue the great giant. The predator entered the shallow water, splashing along behind Brontosaurus, and as he strode along he left footprints in the bottom ooze, three-toed birdlike prints on a magnified scale, prints marked by the deep impressions of the great claws.

Brontosaurus now became aware of her pursuer, so she abandoned her leisurely pace and made for the far shore with all the prodigious energy at her command. Allosaurus also put on a burst of speed, in an attempt to overtake her. His effort, too, was immense, and it brought him abreast of Brontosaurus as she left the water and clambered up the beach. She was trying to reach the forest as he attacked with all of the reptilian fury at his command. His attack was bloody, but largely ineffectual. Her size was her protection: the great predator could tear at her, but he could not bring her down. So they crossed the beach, the one clumsily fleeing as the other attacked time and again, and in that manner they entered the jungle.

Immediately beyond the edge of the forest and behind the barrier of the beach was a large swamp or marsh, an unexpected and welcome haven for the huge sauropod. Tormented and torn by her adversary, she rushed toward the marsh and plunged into it. Far and deep she waded into the marsh, until finally she was submerged to the level of her back. Here she had found her refuge, for Allosaurus was loath to enter such deep water with so treacherous a bottom. His strong, birdlike limbs and his sharply clawed feet were ill-suited to the new situation. He could run across the hard ground, and he could even splash through the shallow water of the embayment, but in the marsh he was at a total disadvantage. So after a few tentative steps into the soft ground that confronted him, he turned and retreated to firmer terrain. He had failed in his attempted predation.

Brontosaurus had succeeded in her escape, but her success was somewhat diminished because her wounds were deep and painful. As she waded into the depths of the swamp the water around her was suffused with the blood from her wounds. Her energy was drained; she rested quietly in the dark, cooling water of the marsh as the afternoon wore on and the sun sank below the horizon.

The Paluxy River, a tributary of the Brazos, is a pleasant stream that wanders through the limestone hills of east-central Texas near the little town of Glen Rose, some sixty miles southwest of Dallas and Fort Worth. There would seem to be nothing especially distinctive about the Paluxy, for to the casual eye it appears as just another of the thousands of minor rivers of North America. But the Paluxy is different.

A few miles from Glen Rose the bed of the river is flat and rocky, and there the banks are low and at times muddy. The rocky riverbed consists of a cream-colored limestone of the early Cretaceous Age; the soft banks are composed of gray-black shales, also of the Cretaceous Age. The limy strata of the riverbed are essentially horizontal, and they continue uninterrupted beneath the overlying shaly banks. Here, as a result of work by the flowing waters, a series of large depressions, known as "potholes," has been revealed.

Potholes are rather common in rocky riverbeds. They may be formed by the water's scooping out soft places in the rock, or quite frequently they are shaped and enlarged by the abrasive action of pebbles or even large stones, washed into depressions and rolled around and around by inexorable pressures from the flowing water. But the potholes in the beds of the Paluxy River are different.

Years ago, people who visited that stretch of the Paluxy River when the water was low noticed that the potholes, as large and as deep as washtubs, were numerous and were aligned. This seemed very strange. River potholes commonly are randomly distributed; why should the potholes of the Paluxy be arranged in a definite pattern?

Back in the years of the Depression, shortly before the outbreak of World War II, the strange potholes of the Paluxy River came to the attention of Roland T. Bird, a paleontologist then associated with the American Museum of Natural History in New York. Bird went to Glen Rose and made his way to the place of the potholes, where he began to examine them in detail. The riverbed was partially dry at the time, and the exposed potholes were partly filled with mud and dust. With a small shovel and a whisk broom Bird began to clear out one of the larger depressions. When he had swept it clean, down to the bare rock, he realized that he was looking not at the usual type of erosional depression, but rather at a gigantic footprint, obviously the footprint of a great sauropod dinosaur. That was evident from its configuration, because instead of being more or less round, as are common potholes, it was somewhat elongated and rather elliptical in shape. One end of the elliptical depression was narrowly rounded, but the opposite end was broad and terminated in three sharp points. Clearly this was a fossilized print made by the hindfoot of one of the greatest of the dinosaurs, a dinosaur exemplified by the skeletons of the gigantic sauropods.

Immediately in front of this track was one of the smaller potholes, a rounded depression indented along its back, or posterior, side. This was clearly the print of a front foot, without any claw marks.[6]

Here were the records of an event that took place in the far-distant past, the imprints made by the feet of a gigantic dinosaur that lived more than a hundred million years ago. But what Bird had uncovered in his preliminary examination was only a small fragment of the record. He could see the trackway continuing along the riverbed and disappearing beneath the shaly riverbank. So he made plans.

The next year he returned to the site with a crew of assistants and proper equipment for a large job of digging. As a result of long weeks of work (during the course of which an extended dike of sandbags was constructed to divert the river from the area of the footprints), a magnificent trackway was exposed, showing the progress of a dinosaurian giant across an ancient mud flat once covered by shallow water. Bird realized that it had been an area of shallow water because nowhere along the trackway was there any evidence of a tail mark. Yet the water evidently was not deep enough to float or partially support the body of the dinosaur, because the footprints were very deep, indicating that the full weight of the animal had been supported by the limbs. The limy sediment in which the footprints were preserved belongs to a geological formation known as the Glen Rose limestone, which from the internal evidence of the rock was deposited in shallow marine waters.

Still another aspect of this record of the past was revealed by the picks and shovels of Bird's crew, for paralleling the trackway of the giant brontosaur, and in places even superimposed on the huge footprints, were the three-toed, birdlike tracks of a gigantic carnivorous dinosaur, of the type represented by skeletons which have been named carnosaurs. It is quite obvious that the big predator had been following the brontosaur closely, because the fact that some of the carnivore footprints were on top of or contained within the sauropod prints shows that they were made after the great giant had traversed the mud flat.[7] Does that indicate that the predator actually was pursuing the sauropod—that we see the record in stone of a hunt that took place almost 200 million years ago? It is intriguing to think so.

SECOND MONTH

The Forest and the Swamp

BRONTOSAURUS spent many days in the swamp, resting and recuperating from the grievous wounds inflicted by the cruel teeth of the carnivore. At first she remained in a state of torpor, silently caressed by the dark waters of the swamp. But within a few days her energy began to revive, so that she was able to wade about from time to time, occasionally browsing upon the aquatic plants that surrounded her. As the days passed, her wounds gradually healed and she became ever more active.

Even so, her movements were slow and ponderous. The depths of the swamp protected her from attack by the constantly aggressive giant carnivores, so there was no reason for her to be constantly alert. There was always vegetation for the eating. Consequently hers was a slowed-down existence.

She was not alone. In the swamp, which was of considerable extent, there were other sauropod dinosaurs like herself, wading and feeding. Some of them had been there when she first plunged into its protecting environs, as she fled from Allosaurus, and at that instant those established inhabitants of the marsh, alarmed by the charge of the attacking carnosaur, had retreated as far as they could from the scene of conflict. Now, when all was quiet, they had returned to where the wounded giant rested, and as time went on they were joined by others of their kind. Indeed, a sauropod herd had gathered together, to rest and occasionally to feed within the haven of the marsh.

It must not be thought that the herd of sauropods remained stationary within the confines of the swamp. There was constant, widespread motion, so that the opaque water, almost choked in many places by abundant vegetation, was virtually never still. Although the plant life of the swamp grew prolifically, the dinosaurs were big, and consequently they required, individually and as a herd, tremendous volumes of food. Their food was bulky, and great quantities of plants were necessary to produce the energy needed to maintain the proper functioning of the gigantic bodies.

The swamp supplied some of that energy. So Brontosaurus moved through the water and across the soft ground, as did her fellows, browsing upon the plants that surrounded her. Within a few moments she would consume the vegetation within reach of her long, ever-active neck, and then she would wade a few paces on to attack a fresh supply of food. At the same time the other members of the sauropod herd were wading and browsing and constantly churning up the water and the mucky bottom of the swamp.

Yet with all of their noisy progress through the swamp water, their feeding was rather desultory, for seemingly it was not the plant life of this watery environment in which they were primarily interested. Beyond the swamp, on the green fringing lowlands, were the plants to which they probably were especially attracted. There were the abundant ferns and club mosses, the tree-ferns, cycads, evergreens, and other trees which formed the main portion of their diet. So it was that the dinosaurs abandoned the protection of the swamp to wander widely through the jungle in search of food.

By and large the herd kept together while they fed and rested, and Brontosaurus, so recently injured by the attack of her predatory enemy, stayed with the herd. The group never moved so fast that she could not keep up with it, and the food in the swamp and in other swamps, or on the higher ground, was restorative, so that bit by bit her recuperation proceeded. The wounds healed further every day, and every day she became less listless and more active. The near-fatal encounter became something of the past.

And so the days passed, partly within swamps and partly out on dry ground, partly feeding and partly resting. Many hours would elapse and perhaps quite a few miles would be covered each day before the various sauropods, large and small, had satisfied their hunger and were ready to rest. The quiet of their resting would last for some hours; then the feeding would begin again. Always for such giants there was the necessity of renewing their energy, ever being drained by the processes of living: the natural fuel consumption known as metabolism.

Thus her mode of life was quite different from that of the giant carnivorous dinosaurs that might at almost any time be seen lurking at the edge of the herd. For them feeding was an intermittent affair. If one of them was lucky enough to pull down a half-grown sauropod, or to finish off an ancient giant weakened by old age, or perhaps to chance upon a half-decayed carcass, his feeding would be voracious and gluttonous. He would gorge himself. Then, thoroughly satiated with high-energy protein, he would endure a fast of many days. It would be a fast during which he would remain torpid, or at least relatively inactive, while his system absorbed the nourishment provided by the huge quantity of meat that he had eaten. So for the carnosaur life was a fluctuating cycle, alternating between periods of frenetic and compelling activity devoted to the hunt, and periods of languid inactivity following a gargantuan feast. It was a decided contrast to the steady, uneventful search for food that was the daily routine of Brontosaurus.

Her browsing was in some respects a slow process, for she had a very small mouth in relation to the bulk of her body. She had to take in food through an orifice hardly larger than the mouth of a rhinoceros to stoke the internal fires of a body twenty times the bulk of a rhinoceros. All of which would make the task seem well-nigh impossible, except for two facts. First, Brontosaurus was a reptile, albeit a special type of reptile, and it is possible, even probable, that her body metabolism was decidedly slower than that of a mammal of comparable size (a modern whale, for example), thereby making her food requirements less than might be supposed. Second, because of her huge size there was a certain degree of conservation of energy, so that the amount of food she ate was less, relative to her bulk, than would be the case for a smaller animal of the same kind.

Consequently, even though her feeding progressed at a comparatively slow rate, it was sufficient. By browsing constantly through long daylight hours, she got the sustenance she needed.

Her method of feeding was simple and direct. Reaching high into the trees that surrounded her or lowering her head to the ground, she would tear leaves or fronds from the primitive tropical plants and swallow them whole. There was little if any chewing, little if any attempt at breaking down or mashing the plant fibers before passing them on to the stomach. It was a matter of biting (or, more properly, cropping) and swallowing, without any intermediate steps to comminute the food. Perhaps, when faced with long leaf-bearing branches or a long frond, she might strip the vegetation from its supporting stem by a sidewise movement of her head. But there were no complicated movements of the jaws, for they were simple and relatively weak structures, suited for the one purpose of cropping plants.

As for swallowing the torn but unbroken leaves, there was the incredibly long passage from mouth to stomach. Down the twenty-foot-long gullet went the green food, wave after wave of raw vegetation, hardly softened by the juices of the mouth. Her esophagus was tough.

Moreover, her stomach was tough. The vegetation lodging in that organ in a very raw condition was probably broken down to be digested by strong gastric juices and bacteria. It may be, too, that she had "stomach-stones," pebbles that had been swallowed from time to time, which by roiling around in the digestive tract helped mechanically to break down plant fibers.

A month had gone by, and the food supply of the

swamps, extensive though they were, and of the surrounding forests was diminishing under the ubiquitous browsing of the gigantic dinosaurs. This was particularly true in the forest, where the reptiles found their favorite food. The ground was trampled, and the most succulent of the ferns were damaged or completely consumed. Trees were stripped of their vegetation. It was time to move on.

Brontosaurus, in company with other members of the herd, led by some instinctive stimulus or beckoned by some remote signal, set out upon a journey to new foraging grounds. Their new promised land was to be seen on the distant horizon, across an embayment of the sea similar to the one Brontosaurus had been crossing when she had been so cruelly attacked a month before. With a remarkable singleness of purpose the group of sauropods moved straight in the direction of their new pastures, through the edge of the somewhat ravaged forest, across a beach, and again into the shallow seawater.

They were all wading through this embayment as Brontosaurus had singly waded through the other bay on the day of her painful adventure. Now she was in the company of others of her kind; perhaps there was some safety in numbers. As the herd made its way across the arm of water, its progress was marked by a broad and confused band of footprints—hundreds upon hundreds of them pressed deeply into the soft, oozy bottom. Yet not all of the footprints were deeply impressed on the floor of the bay. Some of them were small and shallow; when the young sauropods waded in water up to their shoulders, their weight was largely supported by the water, so that their feet touched the bottom lightly. Almost all of the small footprints were impressed within the central part of the wide trail, as the young dinosaurs waded together in a subgroup of their own, protected by a massive, living wall of giants to the right and to the left.

Then the water deepened, so deep that the largest members of the herd were almost floating, with the water up to the tops of their backs. Brontosaurus found that she could no longer walk along on her four feet; her thirty tons of weight had become almost weightless in the deep water. So she began to swim, as the smaller dinosaurs already were swimming, propelling herself forward with sweeping side-to-side movements of her powerful tail. As she swam along with the other members of the herd she would touch bottom in places with her feet. Then she would "pole" herself along with her front feet, like boatmen in shallow rivers and lakes, who propel their boats with long poles pushed against the bottom. Although most of her poling was done with the front feet, she would occasionally put down a powerful hindfoot to give an extra push, or to help her change direction. It was a mixed manner of progression; sometimes wading, sometimes poling, and sometimes swimming free, according to the depth of the water. It sufficed. She made her way straight across the embayment, as did the rest of the herd, despite the varying depths of the water. For those dinosaurs, adapted though they were for walking on powerful limbs across hard ground, the water held no terrors. Indeed the water was friendly, because here in the deeper water of the bay, as in the deep water of swamps, they were immune from attack by powerful predatory dinosaurs.

In that fashion Brontosaurus, along with the other sauropods, reached the far side of the embayment, to enter a strange forest with its new swamps, where food was plentiful.

In a sense life began anew for Brontosaurus and her

fellows in this new habitat. The vegetation was the same, it is true, as the vegetation lately abandoned by the herd, but the plants were fresh. The ground was thickly covered with lush ferns; the trees bore undamaged leaves. So the herd began again to browse in the deep, green shade.

The largest of the sauropods reached high into the trees to crop the most succulent of the leaves, and Brontosaurus joined with them in the delightful feast. They enjoyed particularly the soft, spreading fronds of the tree-ferns, and the long, frondose leaves and the globular fruits of the treelike cycads, the Williamsonias, which grew in abundance within the forest, their scaly trunks crowned by very long leaf-fronds to form great arcades, through which the dinosaurs made their way. To vary their arboreal diet the big dinosaurs would reach down and seize mouthfuls of moist ferns, often shoving aside some of the smaller animals, which were eating the low plants, since they could not reach the high crowns of the larger tree-ferns and Williamsonias.

Not infrequently Brontosaurus would push against a tree and flatten it against the ground, to bring the higher fronds down within easy reach. When she did that, the smaller animals would sometimes rush up to join her at her feast. That method of forest destruction was widely practiced by the larger members of the sauropod herd, so that the vegetation was quickly consumed. For that reason, and because of the constant trampling and browsing of the low plants, the dinosaurs were forced to keep moving as they fed. Their activities wrought considerable damage in the forest, luxuriant as it was.

Whether she was wandering through the forest or wading in the marsh, Brontosaurus made constant and effective use of her long neck. It was almost always in motion, up and down and from one side to the other. To her it was an essential part of her being and of her life, enabling her to reach high and low and to the right and left, not only for feeding but also for the wider contemplation of her environment.

It was like the boom of a great crane, but it was flexible and marvelously strong. From the massive shoulders powerful muscles extended forward on to the enormous neck vertebrae, while down the midline of the neck, lodged in cradlelike, V-shaped vertebral spines, was the great ligament, the *ligamentum nuchae,* attached behind the skull and powered by the muscles from the shoulder region. That was the cable of the boom, serving to raise and lower the head; it also worked with other muscles, which provided lateral movement, making the neck of Brontosaurus a living mechanism of force and significance.

What could have been more perfect and more logical than the neck of Brontosaurus? Its length and its power defined her role as a browser of the upper regions of the forest and very likely as an underwater browser, too. It gave her a measure of versatility in feeding that was in many ways the reason for her success as a herbivorous giant.

Of that she was, of course, quite unaware. The complex employment of her neck to bring her ever-active jaws within reach of food, high in the trees or down beneath the water, was for her a matter of instinctive motion, the result of millions of years of evolution.

So she continued in her search for food, a search made largely possible by the great neck that allowed her to probe and reach, and especially to browse in high, leafy pastures denied the lesser dinosaurian herbivores that shared her world.[8]

In that way Brontosaurus spent her days with the herd, ever browsing and resting, ever moving in the search for more food, ever abandoning the jungle which had been trampled and bruised in favor of new feeding grounds where the trees stood straight and where the ferns on the ground were fresh. Sometimes big predatory dinosaurs would appear to menace the herd and, if possible, to cut down one of the smaller or weaker members that might inadvertently stray from its accustomed place in the center of the group. On the whole, however, it was a peaceful existence, lived through seemingly unending tropical days and nights.

There has been a long debate as to what the giant sauropods ate, and how they ate. It is not easy to relate long-extinct animals to their environment, particularly when one tries to get down to details. It is a matter of putting together small bits of evidence and then making the best possible inferences from such evidence.

Sauropod teeth are simple in form, few in number (compared with the teeth in many reptiles), and not very large in relation to the size of the animal they served. *Apatosaurus*, the genus to which Brontosaurus belonged, had peglike teeth; among other sauropods, *Camarasaurus* had somewhat broadened, spatulate teeth, anteriorly pointed and with sharp, cutting edges, while *Diplodocus* had remarkably slender, pencillike teeth with spaces between them. Almost everybody agrees that these dinosaurs were able to crop vegetation with their simple teeth; the signs of wear on those teeth indicate that they were so used. But the teeth do not show excessive wear or deep striations on their surfaces, as might have resulted if there had been prolonged chewing of plant fibers. So it is logical to think that the food was merely cropped and then swallowed "raw," to be processed in the digestive tract.

There is a trend now to consider the sauropods essentially terrestrial animals, feeding largely on the leaves of trees. Formerly it had been thought that those giant dinosaurs spent most of their time in the water, feeding upon aquatic vegetation. At times some of the modern proponents of the tree-feeding theory view the earlier idea of the sauropods as aquatic browsers with a certain amount of scorn. But why is it necessary to take an either-or position on the matter? Why wouldn't the sauropods have been in and out of the water, like modern elephants, according to the opportunities afforded at the moment? Why wouldn't they have fed both upon trees and upon aquatic vegetation?

There is good evidence to show that the sauropods frequented watery places and often went into deep water. The footprints described in the first chapters are clear proof of this. Moreover, all of the sauropods have highly placed nostrils (in *Diplodocus* they are on the very top of the skull), which strongly implies that those dinosaurs were at least partially aquatic. Elevated nostrils allow an air-breathing animal to breathe while the body and most of the head are submerged. Witness, for example, modern crocodiles and hippos.

Whether the sauropods fed upon trees or upon aquatic plants or both, it does seem evident that they were entirely herbivorous. Certainly there is nothing about their relatively small teeth and jaws to suggest a carnivorous mode of life. It follows that they consumed bulky food, ingested through a comparatively small mouth, all of which required prolonged feeding.

Mention has been made of the conservation of energy, because of the great bulk of the giant sauropods. To state it simply, large animals require less food in relation to their size than do small animals, a function in part of the relation of mass to surface area.

For example, a cube that is 1 foot square will have a surface area of 6 square feet: 1 square foot of area on each of 6 sides. If the unit of mass is assumed to be 1, then the relation of mass to surface area is as 1 to 6. A cube 2 feet square—or twice the size in its linear dimensions—will be 8 times as massive as the 1-foot-square cube (2 × 2 × 2). It will have a surface area of 24 square feet (4 × 6), making a relation of mass to surface area as 8 to 24, or 1 to 3. And a cube 3 feet square will be 27 times as massive as the 1-foot-square cube (3 × 3 × 3), with a surface area of 54 square feet (9 square feet on each of 6 sides). The consequent relation of mass to surface area is as 27 to 54, or 1 to 2. The larger the cube, the less the relative surface area.

Of course animals aren't cubes, but the principle holds. A large animal has less surface area in relation to mass than a small animal, which means that there is less relative area for heat loss. Heat loss is energy loss. Therefore a large animal suffers relatively less energy loss than does a small animal, so it requires relatively less food.

An elephant weighing some 5 tons may require about 600 pounds of natural plant food per day, a ratio of food to body weight of 6 to 100. But those are figures for active, warm-blooded mammals. How can we estimate the food intake for the giant sauropod dinosaurs?

A modern captive Galapagos tortoise weighing 300 pounds consumes per day about 10 pounds of food, consisting of a mixture of "zoo cake," kale, carrots, and salt marsh hay.[9] The ratio of food to body weight is thus about 3.3 to 100, considerably less than for the elephant, even though the tortoise is much smaller than the elephant. It is in part the difference between a "warm-blooded" endotherm and a "cold-blooded" ectotherm.

In a recent paper it has been estimated that a duck-billed dinosaur weighing about 5 tons may have consumed some 450 pounds of plant food per day—a ratio of food to body weight of about 4.5 to 100.[10] On that basis, perhaps a sauropod dinosaur weighing 30 tons six times as much as the duck-bill or hadrosaur—may have eaten as much as a ton of herbage each day. It seems like a lot, but *Apatosaurus* was a monstrously large reptile. Even if the giant sauropod consumed only 3 percent of its weight daily, as in the case of the Galapagos tortoise, the food intake would have amounted to 1,800 pounds, almost a ton. Or suppose, in view of its great size, the relation of food to body weight was on the order of 2 percent. Even so, the sauropod would have eaten 1,200 pounds of food per day. Yet although a ton, or 1,800 pounds, or 1,200 pounds is a lot of plant food to pack in during the course of twenty-four hours, it is considerably less than that required by even the largest of modern land-living mammals: 4 or 3 or 2 percent, as compared with 6 percent.

Thus it would seem that because of their great size the giant dinosaurs probably did not consume as much food daily as might be supposed. Besides, since reptilian metabolism is lower than mammalian metabolism, the probable food intake is reduced even further.

There is the problem, however, of what the levels of metabolism in the dinosaurs might have been. Were they similar to other reptiles? Or, like mammals, did their bodily functions operate at a high level? That complicated question will be explored in a later chapter.

Roland T. Bird, who excavated the dinosaur tracks near Glen Rose, Texas, uncovered another remarkable concentration of footprints on the Davenport ranch in Bandera County, Texas, some two hundred miles southwest of Glen Rose. At Glen Rose, it will be remembered, there was a well-defined trackway made by a large sauropod, and another trackway made by a carnosaur, which obviously was following the herbivorous giant. It makes a dramatic story. At the Bandera locality the story is perhaps not so dramatic—it is not the record of a hunt and of possible conflict—but nonetheless it is fully as impressive. Here the footprints are profuse and crowded together, with many prints superimposed upon others. The tracks face in the same direction and seem to indicate a herd of sauropods crossing a mud flat at a leisurely rate—"like cows going down a country lane," as Bird put it.[11] Moreover, those footprints are of varying sizes, the prints of a group of animals of different ages, with the smaller prints commonly in the center of the trackway.

Was it truly a herd of sauropods traveling together, or do the crowded Bandera footprints represent a fortuitous assemblage of tracks made by solitary individuals crossing a mud flat during a certain lapse of time? We know that basking crocodiles, for instance, if startled, will dash into the water side by side or one behind the other, thus leaving parallel and overlapping trackways within a relatively small area. Those reptiles cannot be considered part of a herd, nor does their behavior, as reflected by the tracks they make, indicate a herd migration.

Nevertheless, the evidence that the Bandera footprints do represent a herd of sauropods is impressive. Bird identified twenty-three trackways making up this aggregation, representing twenty-three individuals. All are headed in the same general direction, their maximum divergence being no more than 25 degrees. As Bird has said, "Of the 23 individuals crossing the area all were headed toward a common objective. This suggests that they passed in a single herd."[12]

In a recent paper by Dr. John Ostrom of Yale University, the Bandera tracks are carefully analyzed, as are various other congregations of dinosaurian footprints. From the accumulated evidence, Ostrom concludes that many dinosaurs were indeed gregarious and traveled in herds. That is certainly a reasonable inference, particularly as applied to the large herbivorous dinosaurs.[13]

If the dinosaurs were headed toward a common objective, as Bird pointed out, what might the objective have been? Of course that is a difficult question, and several answers might be put forward. In that connection an analysis by Dr. Claude Albritton of Southern Methodist University of some dinosaur trackways in the Glen Rose Formation at Lake Eanes, near Comanche, Texas, should be noted. According to Albritton, the dinosaurs were moving with a purpose: "Animals that are browsing do not follow straight paths, nor do they take uniform paces. The trails at Lake Eanes indicate that the dinosaurs were moving straight across and at a steady pace toward destinations which lay beyond the present exposure of mud cracked stratum." Furthermore, he speculated that the straight trails, generally in a single direction, may have indicated that the dinosaurs were taking a shortcut across an exposed tidal flat between two points of land, "a course that was ordinarily blocked or made less attractive by the presence of water."[14]

The story of a herd of dinosaurs crossing an embayment of the sea at low tide thus has a reasonable foundation from the evidence of trackways at various places in the Glen Rose Formation of Texas. Also, those tracks give good evidence as to the methods of locomotion among the sauropod dinosaurs: locomotion by walking and wading, by swimming, and by poling with the feet, especially the front feet, in water deep enough to float the animals.

In that connection it should be noted that the Texas footprints afford some valuable evidence as to how the giant sauropods walked across the land.

Anatomical studies seem to indicate that sauropod limbs were very strong and quite capable of supporting those huge dinosaurs for long periods of time. In addition to the anatomical evidence, the nature of the Glen Rose trackways confirms the position not only that the limbs of the gigantic sauropod dinosaurs were sufficiently strong to support the weight of these reptiles on land, but also that the sauropods were efficient walkers. A crucial feature of those trackways is that they are narrow: in other words, the inner borders of the footprints are within about ten or twelve inches, on each side, of the midline of each trackway. This shows that the four feet of the sauropod were well beneath the body, below the center of gravity for the animal. By inference it indicates that the legs were relatively straight, so that lines of force passed down through the limbs to the feet and to the ground beneath the middle of the body. Such positions for the limbs and feet should not be surprising; they are virtually predicated by the physical problem of sustaining great weight.

The Glen Rose trackways indicate that the sauropods walked quite confidently with firm strides, placing each hindfoot just behind the position that had been occupied by the forefoot of the same side. So it seems evident that the sauropods walked in much the same fashion as modern elephants, which when seen head-on show the left and right feet very close together. Of course this does not rule out swimming and poling in deep water, as shown by the Bandera tracks.

Perhaps the analogies with elephants may be extended. We know that elephants spend long hours in the bush, and that they wander for immense distances across the land. We know also that elephants dearly love the water and will spend hours in and around rivers and lakes. In addition, we know that elephants can wade through deep water—so deep that they are essentially submerged. Could not the sauropods have had similar behavior patterns?

Objections may be made that the sauropods were reptiles and that elephants are mammals, and consequently analogies are not valid. However, the sauropods, and the other dinosaurs as well, were reptiles quite different from those with which we are acquainted. It would seem that during Mesozoic times the sauropods filled the ecological niche, on a magnified scale, that is today filled by the elephants.

The analogy of sauropods and elephants forms the basis for the description of the destruction of trees and other vegetation by the gigantic reptiles. Modern elephants wreak great damage in the bush as they browse upon trees. It seems logical to suppose that the sauropod dinosaurs may have behaved in a somewhat similar fashion.

The problem of the long neck in the sauropod dinosaurs, of prime importance in their search for food, is more complex than might be supposed. There is much more to a very long neck than merely stretching out a short neck. The long neck is not only a long lever, requiring a strong fulcrum and very considerable power to operate it; it is, of course, a flexible lever. The modern giraffe is provided with a long neck for treetop browsing, the result of long ages of evolutionary development. It works, but its flexibility is limited because the giraffe, like other mammals, has only seven vertebrae in the neck. The neck vertebrae in the giraffe are quite long, with the result that the neck cannot be bent sharply. The sauropod dinosaurs had an advantage in that respect, because there were fourteen vertebrae in their necks. So the dinosaur's neck had a degree of flexibility never attained by the giraffe.

All but the first of the sauropod neck vertebrae were greatly enlarged for the attachment of strong muscles. These muscles, in conjunction with the great nuchal ligament, gave the sauropod neck its tremendous strength. It had to be strong to perform the functions for which it was adapted.

Once again a comparison will be made between the sauropod dinosaurs and the greatest land animal of our day, the elephant. Elephants have very short necks—so short that elephants can move their heads only through limited arcs. But elephants have their remarkable trunks—long, very flexible, and remarkably powerful. Elephants can do in a different manner with their trunks what sauropod dinosaurs presumably did with their long necks to minister to their many needs.

Attention so far has been restricted to the mechanical problems of the sauropod neck. But there are other problems as well. One has to do with blood circulation; another, with breathing. Perhaps the comparison between the sauropod dinosaur and the modern giraffe with its extraordinarily long neck may be extended to those two particular problems.

The giraffe has very high blood pressure as the blood supply leaves the heart and enters the carotid artery. Such high pressure—much higher than in any other mammal—is necessary to drive the blood up to the head, far above the heart. Why, then, doesn't the giraffe suffer from blackouts caused by the change in blood pressure from the arteries to the brain? Recent studies have shown that the arteries entering the giraffe brain divide into networks of fine vessels, the *rete mirabile.* Although there is as yet no definite proof, it seems possible that the *rete mirabile* may regulate blood pressure so that in the brain such pressure is more or less equivalent to what it is in other mammals.[15] One wonders if a similar mechanism was present in the giant sauropod dinosaur, whose head was far higher above the heart than is the giraffe's. That is a matter for speculation.

As for breathing, there is the problem of a very long trachea, or windpipe. In the giraffe the trachea is longer than five feet. All of that length is in effect dead air space where there is no gas exchange. In the sauropod dinosaur that space was at least fifteen feet long. Such a length of dead air would effectively block the proper ventilation of the lungs or the exhalation of carbon dioxide without some mechanism for overcoming it. In the giraffe the solution to the problem is hyperventilation: breathing rapidly and deeply. Is it possible, then, that the sauropods were of necessity forced to breathe much more rapidly than other reptiles? Again, that is a matter for speculation.

THIRD MONTH

The Land of Giants

IT WOULD SEEM that the days within swamp and jungle followed one after the other in monotonous succession. They were, on the whole, peaceful days spent in a tranquil environment. And they were peaceful because Brontosaurus was, like her fellow sauropods, a giant.

She had little to fear. Of course there might be occasional threats of attack, or even actual attacks, such as the one she had experienced some two months ago. But direct attacks by the awesome predatory dinosaurs were uncommon as far as she was concerned. She was no longer alone, and the great reptilian carnivores, terrifying though they might be, generally avoided the congregation of giants and directed their destructive attentions toward the smaller and weaker members of the herd, those juveniles or aged and infirm sauropods that might stray from the protection of the group. Seldom did they attack a full-grown, vigorous sauropod that was consorting with others of its kind. There was safety in size and safety in numbers. That was why Brontosaurus, moving with the other sauropod dinosaurs, was a gentle giant: giants can afford to be gentle.

During the long, dreamlike days of wandering and feeding, Brontosaurus had many encounters and confrontations, mostly uneventful. Yet they punctuated the usual even tenor of her quest for food. She saw much of the other members of her herd with whom she moved, often shoulder to shoulder, so she was accustomed to them. But she saw other sauropod dinosaurs, too, because the tropical world in which she lived was not an exclusive feeding ground for sauropods of her ilk—the particular giants which may be called brontosaurs. There were many and varied sauropod giants to share the land. As today there are diverse great whales in the seas, so in the days when Brontosaurus lived there were differing great sauropods on the continents.

Quite frequently she and her companions would come face to face with a herd of camarasaurs (*Camarasaurus*), sauropod close-cousins that also wandered among the endless green mansions of the Mesozoic world. Camarasaurs resembled Brontosaurus in many respects. There was, however, some difference in size: the largest camarasaurs were not quite as massive as the largest brontosaurs.

There was also a difference in general profile—a difference that was, perhaps, apparent and meaningful to the two types of sauropods during their mutual encounters. Whereas Brontosaurus and her fellows possessed forelimbs that were rather short compared with the massive hindlimbs, causing the shoulders to be somewhat lower than the hip region, the limbs of the camarasaurs were more nearly equal in length, so that the line of the back was horizontal, rather than sloping from hips to shoulder. It may be that this difference constituted a ready recognition signal, enabling the two types of sauropod dinosaurs to distinguish friends and strangers, even at some distance.

Finally there was a qualitative difference: the camarasaurs had larger, more closely set, and sharper teeth than the brontosaurs. Perhaps they were the more efficient browsers, or perhaps they ate different plants from those that formed the basis of the brontosaur diet.

However, the differences between the brontosaurs and the camarasaurs had little effect on their relationship; they tolerated each other. Brontosaurus moved among the lesser camarasaurs much as she moved among her own kind, in concordance and with no feelings of hostility or apprehension. That feeling of tolerance toward the camarasaurs was enhanced because those dinosaurs were smaller than Brontosaurus and her fellows. She was impressed by height, and when she could look down at some of her neighbors and relatives she was not impressed.

So she was impressed, and very much so, when she and her band would occasionally encounter a group of sauropods that we now know as *Barosaurus*. They were giants as bulky as the brontosaurs; in addition, they possessed wonderfully elongated and massive necks that enabled them to lift their heads high into the forest to browse at levels beyond the reach of the brontosaurs. Thus the barosaurs could look *down* at Brontosaurus and her accompanying herd, which was a decided advantage for those looking down and a disadvantage for those that had to look up. Brontosaurus was rather wary whenever she encountered a barosaur; perhaps she was not intimidated by her long-necked cousins, but she did feel some apprehension in the presence of a group of barosaurs. She maintained a little distance between herself and the barosaurs, if possible, yet she was willing to join them in a spacious feeding ground.

One day as Brontosaurus was browsing through the jungle, a little in advance of the other members of the herd, she met a giant of such overwhelming proportions that she, for once, was dwarfed by comparison with this intruder. It was a brachiosaur (*Brachiosaurus*), twice as massive as Brontosaurus, with long forelimbs and high shoulders and a long, heavy neck, enabling him to reach up forty feet into the trees to browse. And the head of *Brachiosaurus*, carried so high on his long neck, was distinctively different from the head of Brontosaurus, for it was crowned with a narrow, longitudinal, rounded crest, within which, it would seem, the nostrils were lodged. In spite of his size the brachiosaur was in no way menacing; he ignored Brontosaurus as giants often do when they encounter lesser creatures—indeed, much as Brontosaurus had ignored the camarasaurs. Brontosaurus nevertheless was intimidated—more by the great bulk and height of this dinosaur than by his manner. So she backed off and turned in another direction.

Now and again she would meet a group of very slender and elongated sauropods which we call *Diplodocus*. The largest of the diplodocines might be more than ninety feet long; although longer than Brontosaurus, those dinosaurs were not nearly so massive, because of their very slender build. Indeed, a big *Diplodocus* was only about a third of the weight of Brontosaurus, so she felt no unease at being in the midst of a group of such dinosaurs. Furthermore, the diplodocines seemed to frequent deep water and swamps more than Brontosaurus and her kind, so that her encounters with those attenuated sauropods were less frequent than her encounters with the camarasaurs. *Diplodocus* had nostrils on the very top of the head—commonly a feature of aquatic, air-breathing animals, perhaps permitting that dinosaur, like *Brachiosaurus*, to thrust the nostrils above the surface of the water for breathing, while otherwise submerged. The teeth were very weak, as has been noted, suggesting that *Diplodocus* fed upon soft, aquatic plants, or perhaps upon thin-shelled mollusks.

There were other browsers, too, in the jungle, and Brontosaurus met them almost daily. She stood aloof from the stegosaurs, bizarre-looking dinosaurs of moderately large size, some twenty feet or so in length. Stegosaurs had very short, heavy forelimbs, so that the shoulders were quite low compared to the elevated hip region. The head was ridiculously small, and the jaws, forming a sort of beak in the front part of the mouth, were provided on their sides with small, weak teeth. It seems obvious that these dinosaurs fed peacefully on the tropical vegetation around them. Particularly notable in *Stegosaurus* was the double row of large, upstanding, and alternating triangular plates, running down the back—from near the back of the head to a position well down on the tail. And behind the plates, near the tip of the tail, were two pairs of long spikes. Brontosaurus did not particularly fear the stegosaurs—they were basically harmless creatures—but she was wary of those sharp tail spikes. The stegosaurs must have been dim-witted, even in reptilian terms, because the brain was remarkably small for such a large reptile, and it is likely that the reactions and movements of *Stegosaurus* were highly instinctive and automatic. Thus one of these strange dinosaurs might lash out with a spike-studded tail at any large dinosaur of a different kind that had the temerity to venture close.

As for the gigantic carnivorous dinosaurs, always lurking in the vicinity of the brontosaur herd, many were allosaurs, like Allosaurus which had attacked Brontosaurus. But there was another great carnivore: *Ceratosaurus*, a somewhat smaller cousin of *Allosaurus*, distinguished by a horn on its nose. It, too, was a potential threat to the wandering sauropods.

Such were the dinosaurian giants that shared the land with Brontosaurus. Most of them were neither enemies nor competitors; they wandered together through the jungles of that distant past in somewhat the same manner as the hosts of antelopes on the African veldt do today. Brontosaurus and her companions, the camarasaurs, barosaurs, brachiosaurs, and diplodocines, and the stegosaurs, too, formed on a grand scale the ancient counterparts of various antelopes, such as the wildebeests, hartebeests, kudus, elands, and impalas that the traveler can see today in such still-impressive numbers, grazing and browsing across the plains of southern and eastern Africa.

Yet not all of the dinosaurs encountered by Brontosaurus were giants. There were the camptosaurs—vegetarians of very modest dimensions. *Camptosaurus*, a bipedal browser, perhaps some twelve feet or so in length, inhabited the landscape in substantial numbers. The camptosaurs commonly walked in a semierect posture on strong hindlimbs; the forelimbs were short and the "hands" were small. Like a kangaroo, this dinosaur could also walk on all four feet, but perhaps most of the time it traveled by using its powerful hindlimbs. The head of

Camptosaurus resembled in many respects the head of *Stegosaurus*, but the jaws were more generously supplied with teeth than the stegosaurian jaws. The teeth, limited to the sides of the jaws, clearly were adapted for cropping vegetation; in front the upper and lower jaws took the form of a sharp beak. These inoffensive plant-eaters wandered in and out among the giants and probably were little noticed by Brontosaurus and her kin.

There were even smaller dinosaurs in the forest. *Ornitholestes*, lightly built and scarcely six feet in length, was an agile little carnivore, running on long, slender, birdlike hindlimbs. Unlike its biped neighbor, *Camptosaurus*, the attenuated forelimbs of *Ornitholestes* terminated in facile, long-fingered "hands," each hand with three clawed digits. They were very useful for grasping little lizards, insects, and any other small game, but they could not be used for locomotion. The neck was supple and the lightly constructed head had jaws armed with razor-sharp little teeth. With spry leaps and darting runs through the ferns, this active hunter of small game kept clear of the great, ponderous feet of *Brontosaurus*.

There were other moderate-sized and small inhabitants of the land which were not dinosaurs. Lizards were plentiful and furnished a living, in part, for *Ornitholestes*. There were turtles and crocodiles—some of fair size, some quite tiny. And there were rhynchocephalians, rather like large lizards in appearance, but quite different in their origins and taxonomic relationships. The ancestors of the rhynchocephalians were not far removed from the distant ancestors of the dinosaurs themselves. But the rhynchocephalians had followed a very distinct line of development; for example, they had sharply beaked jaws and specialized teeth fused to the inner sides of the jaws, rather than being set in sockets. The rhynchocephalian contemporaneous with Brontosaurus is known as *Opisthias*. Today this line survives in the form of the tuatara, *Sphenodon*, living in a restricted habitat in New Zealand.

Reptiles not only filled the land that Brontosaurus wandered across with her companions, but also had established a domain in the sky. Pterosaurs, or flying reptiles, flapped and glided above the treetops as well as through open glades in the jungle; they skimmed the marshes and lakes and flew out over the sea, where breaking waves lapped the white beaches. They were searching for their food: insects in the forest and little fishes in the water.

These flying reptiles were not very large, perhaps the size of robins or crows. They flew with leathery wings, each wing supported by a long finger—the fourth digit—which formed its leading edge. The first three fingers formed little hooks near the wrist, evidently useful for hanging on branches or on the rough faces of rocky cliffs. The early aeronauts were well adapted for flying; the wings were long and strong. In contrast to the forelimb wings, the hindlimbs were small and weak, like those of a bat. Also, like a bat, it would seem that the bodies of these reptiles were covered with very unreptilian fur, to insulate them and to conserve the energy produced by the heat of the body for the important function of flying.[16] The long, slender jaws were provided with sharp, needlelike teeth.

Here were reptiles that had invaded a new environment—an environment from which backboned animals hitherto had been almost excluded. But the freedom of the air was not long to be monopolized by the flying reptiles. Already the first birds, known as *Archaeopteryx*, had made an appearance upon the evolutionary stage, inhabiting forests that grew on tropical islands and peninsulas in the coralline seas of what is now western Europe. *Archaeopteryx*, fully feathered, with feathers radiating out from a long, bony tail, and with teeth inherited from its immediate reptilian ancestors (which probably were small dinosaurs) still present in the jaws, may have lived in the forests that were home to Brontosaurus. Certainly this early bird was a contemporary of the close cousins of Brontosaurus then inhabiting the European continental region.[17]

As for other backboned animals, there were frogs in the ponds, as well as heavily scaled lungfishes, swimming sluggishly through the shallow, muddy waters, often coming to the surface to gulp air, when the water became too thick or too foul for their gills.

Insects were everywhere, as were other animals without backbones—scorpions and worms on the forest floor and mollusks in the lakes and rivers. And of course the sea was the home of abundant life—all sorts of marine invertebrates and fishes.

That was the world of Brontosaurus, a world dominated by giants of various sorts, but a world in which there were many smaller creatures as well. Even though the lesser forms of life probably far outnumbered the giants, the giants set a scale of living particularly characteristic of that distant day. Life was lived abundantly and in large dimensions.

There are decided advantages to being a giant; one is the conservation of energy, already discussed. Others will be considered in following chapters. Here attention is focused on the boon of security: freedom from attack, freedom from being constantly alert to danger from other animals. It is a situation greatly to be desired, and it is generally one of the benefits of large size, especially of very large size.

In the world as we know it, the elephant, the largest living land mammal, is virtually immune to attack (except from humans, the most dangerous of all animals). In wild Africa elephants go where they will, and lions in the path mean nothing to them. If they choose to walk where there are lions, then the lions, legendary rulers of the forest, must move out of the way. Generally it is a peaceful maneuver; the elephants move majestically along their chosen course, and any lions that may be on the line of the elephant march will slink off to one side, perhaps expressing their feelings with a series of snarls or growls. Sometimes, if elephants are annoyed, they will chase lions. According to one eyewitness account, the pursued lions "ran low to the ground giving vent to furious growls. The elephants did not stop until the lions ran up a tree, where they were finally safe."[18] The giants express their dominance in no uncertain terms.

Lions will attack baby elephants separated from the herd, but if an attacking lion is caught by the mother elephant, retribution may be sudden and terrible. Of course elephants are dangerous when thus threatened or bothered. They value their rights and do not like to be crowded. Yet on the whole elephants are peaceful animals (as evidenced by the ease with which they have been used by man through the centuries).

By analogy it is reasonable to suppose that the huge sauropods, like the elephants the largest creatures in their environment, and like the elephants browsers of plants, were similarly gentle and inoffensive—if left alone. There was no particular reason for them to be otherwise. Indeed, the sauropods may have been inoffensive to a degree never equaled by elephants. They were huge, but they had no true weapons of offense comparable to the elephants' tusks. For them size, and size alone, probably was their protection.

Perhaps, in view of the foregoing considerations, some objections will be made to the story of hunted and hunter set forth in the first chapter. Such attacks probably did occasionally take place, but they were probably infrequent and unsuccessful. The lion is intelligent enough to realize the odds against it when it looks longingly at a young elephant. The carnivorous dinosaur, with a reptilian brain, was probably less able to weigh odds in advance.

From what we know of modern reptiles, we can conclude that a predatory attack is a matter of instinct, coupled with a certain amount of cunning. For example, it is known that crocodiles and their kin capture their prey by stealth: by approaching their victims quietly in the water, or by ambushing mammals that may come along a game trail to a water hole. However, the reaction of the crocodile or alligator to its intended victim is in essence an automatic response to certain stimuli. If the animal to which the reptile has turned its attention appears unduly large, it will not be molested. In his excellent book *The Last of the Ruling Reptiles,* devoted to modern crocodilians, Wilfred T. Neill states, "Apparently the reptile reacts not to the bulk but simply to the vertical height of another animal. The available data suggest that alligator attacks upon man, when not directed toward children, have been directed toward adults who were recumbent, kneeling, crouching, sitting, or swimming. There is no good evidence that an alligator has attempted to prey upon an adult person who was standing erect."[19] The author then goes on to suggest that perhaps the long necks of the sauropod dinosaurs may have functioned to deter attacks by giant predators, especially when a sauropod raised its head high in the air. If that is so, then it could be that Brontosaurus, wading across the shallow embayment, with part of her height hidden by the water, may not have looked sufficiently tall to deter the attack by Allosaurus. We can only surmise.

It may be that large sauropod dinosaurs occasionally were subjected to concerted simultaneous attacks from several giant predators, the dinosaurian equivalents of modern wolf packs. Even so, the thesis holds that the great giants among the dinosaurs were on the whole immune to attack by predatory dinosaurs. The predators probably concentrated their attention on dinosaurs of less than maximum size.

The characterization of stegosaurs as "dim-witted" may be unfair, but it probably is not. None of the dinosaurs were especially intelligent, for their actions and reactions were ruled by relatively small, primitive reptilian brains. The brain of *Stegosaurus*, however, was especially tiny for an animal of its bulk. *Stegosaurus* was a reptile having a live weight of about two tons, perhaps more for the larger individuals. Yet its brain weighed only a few ounces and was shaped something like a bent cylinder, about four inches in length and no more than an inch and a half in diameter. It was an exceedingly simple brain and it evidently worked in a simple fashion. One might say that *Stegosaurus* was in effect a walking automaton.

The picture here presented of great sauropod dinosaurs peacefully mingling with other herbivorous dinosaurs again is based upon analogies with modern herbivorous mammals, particularly those of Africa. It is a common sight there to see all manner of plant-eaters mingling on the veldt. Herds of elephants, by reason of their size and their crashing progress across the landscape, may keep other herbivores at a little distance, but usually the distance is not great. All of the herbivores tend to mingle in profusion, along with the elephants, giraffes and zebras, wildebeests, and a host of other antelope feeding side by side. When forage is plentiful, there is room for all.

Certainly the forage must have been plentiful during the years of the dinosaurs. Those great reptiles lived in a tropical world of botanical exuberance. With an abundance of plant life around them and with generally benign tropical temperatures through the days, the months, and the years, there was good reason for many of the dinosaurs to become giants. It was a good way of life, especially for reptiles.

For the greatest of the giants, the huge sauropods, it was a supremely successful manner of life, if the abundance of their fossils in middle and late Mesozoic sedimentary rocks has any significance. Those were the years when the world was a place for giants and gentle giants reigned supreme.

particularly felt by Brontosaurus and her fellows. The rising temperature of the air was being gradually transmitted into the massive bodies of the dinosaurs; at the same time the expanses of leathery skin on the great reptiles were insufficient to allow the trapped heat to escape rapidly. Brontosaurus began to feel uncomfortably warm, and this made her uneasy. Her feeling was not that of distress, rather of discomfort.

She needed relief, and nearby were the enticing waters of a swamp; toward that watery haven she now directed her steps, followed by some of the other herd members. She reached the edge of the swamp and pushed her way into it, venturing into ever deeper water until at last she was almost submerged. The water of the swamp, markedly cooler than the air above it, assuaged her somewhat overheated body so that the discomfort she had felt gradually began to disappear. It was the restorative balm that she needed, and quite obviously what many of her fellows needed as well, because within a short while she was in the midst of a considerable number of wading and wallowing sauropods. By now the youngsters, too, had joined the amphibious activities of the sauropod herd, even though, since they could lose heat at a relatively rapid rate, they felt less need for the cooling effect of the water than their huge elders. For them instinct was strong, and it directed them to stay with the herd, wherever it might go.

So Brontosaurus again found herself in the rather deep waters of a swamp, feeding upon the aquatic vegetation that grew there. Immersed in the water, she was protected from the debilitating heat of the humid air. In comparative comfort she waded back and forth during the afternoon, as did the other big dinosaurs, followed by the smaller members of the herd, which were often forced to swim or to pole themselves along through the water. In this manner Brontosaurus and her sauropod companions spent several leisurely hours during the heat of the day.

Finally, however, the sun began its descent toward the western horizon, and slowly the air began to cool. The time had come for the herd to leave the swamp, to engage in some late afternoon browsing among the trees, so they came out, singly and in little groups, to seek the leaves of the forest to which they were so strongly addicted. Some of the smaller sauropods were among the first to leave the water, but in short order they were joined by the other members of the herd, including Brontosaurus. By now she was nicely cooled as a result of her hours in the deep waters of the swamp, and she felt quite comfortable.

Soon she made her way toward a rather thick stand of tree-ferns and Williamsonias, not far from the edge of the water. Here she could indulge in a final bit of feeding before darkness. But her treetop browsing was not to be entirely uninterrupted, because as she thrust her head into the top of a succulent tree-fern she was mildly surprised by some other arboreal feeders. There, clinging to the branches a few inches in front of her jaws, were several small, furry, mouse-gray animals, indeed no larger than mice, but rather different in appearance. These were pantotheres, some of the very early mammals that lived during the age of dinosaurs. They certainly were not very large. In fact, they were so remarkably small that the sauropod head looming up in front of them was, by comparison, gargantuan. Of course the sudden appearance of such a frightening apparition in their midst caused instant and complete alarm. With frightened squeaks the pantotheres dispersed, scrambling and scurrying through the treetops and along the branches in an impressive show of lilliputian speed. Their one aim was to get as far away from that terrifying head as possible.

Perhaps it was well that they did hurry to get out of the way; the browsing habits of Brontosaurus were ungainly, to say the least, and there was every possibility that some of the pantotheres might have been crushed or torn by the probing jaws. But the browsing brontosaur was not an enemy; she had no interest in the pantotheres. It was only the primitive, high-level herbage that occupied her attention.

These tiny mammals, scampering along the limbs in terror to escape her jaws, were the ancient progenitors of animals that eventually were to inherit the earth and to reign supreme after the demise of the dinosaurs. In the days of Brontosaurus they were anything but impressive, for they were very small and weak, with pointed noses and minuscule teeth, fit only for chewing insects and other small game, or vegetation, and with slender little limbs and clawed feet. Nevertheless the insignificant creatures had within their small bodies great potentialities. They were warm-blooded and active, so they were able to escape from the danger threatening them with almost lightninglike rapidity. They were covered with fur to protect them from their environment and to conserve their body heat. Their little teeth were nicely arranged to mesh together, so that they could chop their food into small pieces, thus making it all the more quickly available to be converted into energy. And they had a new kind of brain: not a very big brain, it is true, yet a brain in which the cerebrum, the "gray matter," was becoming dominant over its other parts, making it much more efficient than the simple reptilian brain.

The fact that the pantotheres could move so quickly out of the way of Brontosaurus's jaws (and incidentally could sustain their quick movements for a considerable period of time) was significant. It pointed to a new plan for living.

All of that made little, if any, impression upon Brontosaurus. She went ahead with her browsing, as if the pantotheres had never been in that particular treetop. She continued feeding for some time, while the earth turned and the sun sank ever lower. As the sun went down, the air, once so hot and humid, became progressively cooler, a coolness that eventually was felt by Brontosaurus. Her great body lost heat slowly, just as it gained heat slowly, but after a considerable lapse of time the heat loss did take place, and now she felt slightly uncomfortable because of the low temperature of her environment. Night was approaching, and with the night came a pervading coolness even in this tropical climate. Now the waters of the swamp, which had proved to be an escape from the heat, beckoned to Brontosaurus as a protection against the cool air of the night.

The small lizards, feeding along the ground near her feet, began their retreat beneath the earth's surface as the cooler air rolled across the land. For Brontosaurus there was no underground place to hide; her refuge was in the swamp. She accordingly turned and once more made her way from the jungle to the swamp, where she was to spend the night. With her went the other members of the sauropod herd. Here in the water, which was now warmer than the air, the big dinosaurs could rest through the hours of darkness. It would be for them a time of rest, and little else; there would be no aquatic browsing during the night. Rather it would be a time of stillness and sleep, with gigantic bodies remaining motionless in the quiet water.

So the night passed, hour after hour, with the noises of the jungle filling the nocturnal air, but with nothing to disturb Brontosaurus and her fellows as they rested in the dark water. For them the night was not long; it was merely a pause in their daily routine of feeding. Time was of little consequence to these reptiles.

Yet time did pass, and with the turning of the earth the sun at last appeared again on the eastern horizon. The first rays of light brought activity once more to the herd of sauropods. The light was their signal instinctively to begin once again the wandering and feeding which occupied so many of their daylight hours. Slowly they began to stir and to wade toward the firm ground at the edge of the swamp. With Brontosaurus in their midst the herd came out of the water to enter the jungle with its firm footing. Once in the jungle, they turned toward the bright light that marked another opening or glade. With steady tread the herd pushed through the trees, still shrouded in a sort of half darkness, toward the sunlit glade.

As on the previous morning, Brontosaurus at the edge of the clearing surveyed it briefly before entering into the full light of the sun. And as on the previous morning, the glade was soon filled with the gigantic forms of the sauropods, standing quietly to absorb the delicious warmth of the morning sun.

The story set forth in this chapter has been written on the assumption that the dinosaurs to some extent were affected by, and responded to, the temperatures of their immediate environment. According to that assumption, they did not have strong internal controls over their body temperatures. But perhaps such controls did exist in the dinosaurs, which raises some questions. Were these reptiles *ectothermic* like modern reptiles, with their internal body temperatures relying entirely on their environmental surroundings, or were they *endothermic* like many modern mammals, with internal physiological mechanisms for the control of body temperatures? Or were they perhaps *heterothermic*—somewhat between the ectothermic and endothermic conditions? Those are questions probably never to be solved to our satisfaction. Unfortunately we were not there to observe the dinosaurs at first hand; we must rely upon secondary evidence and upon comparisons with animals we know. Let us look at the evidence and the arguments.

The nearest relatives to the dinosaurs among modern reptiles are the crocodilians: the crocodiles, alligators, and the gavial. Therefore it seems logical to study crocodilian physiology and behavior as clues to dinosaurian behavior. However, there are other relatives of the dinosaurs living today: the birds. So it is equally logical to study bird physiology and behavior, or perhaps some aspects of them, as clues to conditions among the dinosaurs.

The crocodilians are ectothermic animals, as are other reptiles in the world as we know it. Their lives are very definitely governed by the temperatures of the environments in which they live. They are large reptiles, limited to the tropics and subtropics of the earth. In such environments the crocodilians have been very successful indeed, persisting in great numbers and considerable variety from the days when dinosaurs ruled the earth to the present time. It is true that crocodilians are now diminishing, but that is due to the quite unnatural presence of modern humans armed with powerful guns.

Crocodilians are confined more or less to the equatorial belt between the tropics of Cancer and Capricorn because it is not possible for them to go deep underground to escape cold winters. The various crocodilians frequently do have dens along the banks of rivers, but such dens, although offering some protection against the elements, cannot be sufficiently deep to afford an effective retreat against truly cold weather. The crocodilians are therefore creatures of the tropics and subtropics because of their size. One would suppose that the huge dinosaurs, much too large to escape environmental temperature changes except by going into the water or into the shade of trees, likewise were confined to tropical and subtropical regions—if they were ectothermic. (It should be pointed out that some small lizards and snakes are able to live in very cold regions because they can burrow deep into the ground to survive the winters. It is not the winter temperatures when they are protected that determine their distributions, but rather the summer temperatures, when they are out and exposed to the elements.)

The crocodiles and alligators and their kin adapt their daily and seasonal activities to the vagaries of the thermometer. They come out in the morning to bask in the sun (as do many reptiles), thereby bringing their body temperatures to an optimum. When the optimum is reached, they then retreat to the shade or to the water. For the remainder of the day they seek the sun or the shade, according to the need for regulating their internal temperatures.

It was my privilege, quite a number of years ago, to assist in some experiments on temperature tolerances in the American alligator. I was working with two outstanding herpetologists and good friends, and our work was done in Florida in the summer.[20] We had a graded series of alligators, ranging from very small to rather large individuals, and we subjected those reluctant and not very cooperative reptiles to a series of tests in the sun, in the shade, in water, and in a constant-temperature chamber.

We found that the lethal temperature for the American alligator is 38 degrees Centigrade, which is approximately 100 degrees Fahrenheit. When the body temperature of an alligator reaches 38 degrees the reptile is in trouble, and if the temperature stays at that level for very long, the alligator will die. But the alligator likes a temperature just a few degrees below the maximum—say at about 34 degrees Centigrade. At that temperature the alligator functions most efficiently and is most active. The reptile will govern its behavior during the day to maintain a temperature more or less at that level. If the body temperature drops appreciably lower, then the alligator begins to get sluggish.

We learned also, as we had expected, that when exposed to the sun the small alligators absorbed heat much more rapidly than the large alligators, and when they were removed to the shade they lost heat equally fast. That was, of course, a function of body mass; it takes more time to heat a large body than a small one, and a large body loses heat more slowly than a small one. The difference in that respect between the smallest and the largest alligators was quite striking.[21]

That brings up an interesting speculation. Extrapolating

from the alligator, is it not possible that the gigantic dinosaurs had rather constant body temperatures and therefore shared some of the attributes of the warm-blooded birds and mammals, without actually being endothermic? In other words, wouldn't a thirty-ton dinosaur, or even a smaller one, have required a long time to raise the body temperature one degree when standing in the sun, and wouldn't such a reptile have required an equally long time to lose a degree of body heat when standing in the shade? Therefore, is it not possible that the big dinosaurs were constantly at about optimum temperatures, but still sensitive to heat and cold, showing little of the up-and-down variations in body temperatures that characterize the reptiles known to us? And if that was the case, wouldn't those great reptiles have been in some respects about as efficient as mammals, enjoying a high rate of metabolism? Perhaps that explains the success of the giant dinosaurs through millions of years of earth history.

But what about the smaller dinosaurs—the young of the giants, and those dinosaurs that never attained gigantic stature? If they were ectothermic, one would have to assume that their temperatures did fluctuate, like the temperatures in modern reptiles, so that like their present-day relatives they were not as efficient as endotherms of comparable size. Yet they managed quite well; the fossil record attests to this.

To look at the other side of the coin, what is the evidence that the dinosaurs might have been independently endothermic? In the first place, there are the remarkable similarities in many anatomical features that link dinosaurs and birds. Indeed, there is persuasive evidence to indicate that birds arose from dinosaurian ancestors.[22] But does that necessarily mean that the dinosaurs were endothermic? Perhaps endothermy in birds developed after their separation from their dinosaurian forebears.

In this connection it should be said that to date there are no indications that the dinosaurs had a feathery or hairy body covering to conserve heat, as might be expected in an endothermic animal. Numerous specimens of fossilized dinosaur skin have been found, and they all indicate a leathery body covering, such as is found in large modern reptiles.

Beyond the dinosaur-bird relationship several arguments have been advanced in favor of the possible presence of endothermy, or "warm-bloodedness," as a dinosaurian characteristic. One of them has to do with the "erect" posture so obviously characteristic of the dinosaurs, as contrasted with the "sprawling" posture seen in modern reptiles. This supposed difference between dinosaurs and most other reptiles, which has been much emphasized by Ostrom and by Bakker,[23] is perhaps not as definitive as those authors indicate. In the first place, an "erect" posture (with the four legs drawn in beneath the body) was a necessity among the large dinosaurs in order to support their great weight. It was probably a factor of size rather than of physiology; "sprawling" was physically impracticable for animals weighing many tons. Moreover, some modern reptiles are not necessarily "sprawling" animals. Crocodilians commonly walk and run with the legs drawn in beneath the body, and the same is true for various lizards—for example, the gigantic Komodo lizard, which is an active predator on mammals as large as deer and pigs.

Although it has been argued that all ectothermic animals were and are "sprawling," and all endothermic animals "erect," Bennett and Dalzell have shown that this is not true: "Alligators possess a semi-erect stance and have four-chambered hearts, diaphragms, and elaborate lungs, but their metabolic rates are indistinguishable from those of other reptiles."[24] Furthermore, various modern endothermic mammals have postures that are more "sprawling" than "erect."

Continuing the argument that links endothermism with an "erect" posture, Bakker has claimed that the large dinosaurs were rapid runners, and that endothermism was necessary for such activity.[25] But in a recent analysis of trackways made by large dinosaurs, Alexander comes to the conclusion that these ancient reptiles were more likely lumbering animals than "lively runners as shown in some recent restorations."[26]

A very ingenious argument has been advanced to the effect that the ratios of predators to prey among the dinosaurs are more closely comparable to what holds among modern mammals than those ratios characteristic of modern reptiles. That, however, is a very tenuous thesis—so much is dependent upon the accidents of fossilization and preservation. How do we know that the ratios in the fossil specimens represent anything like the true conditions existing when the fossils were living animals?[27]

Perhaps one of the most persuasive arguments in favor of dinosaurian endothermism is in the histology, the microscopic structure, of the bones. The matter has been carefully investigated by various authors, notably by Ricqlès[28] and Enlow and Brown.[29] Ricqlès has suggested that because of the dense Haversian system—the arrangement of canals within bones for the containment of blood vessels, lymph vessels, nerves, and marrow—in dinosaur bones, which resembles the conditions in birds and mammals rather than in characteristic reptiles, the dinosaurs may have benefited from a high rate of metabolism. That evidence has been used by some authors as almost proof positive of endothermy in the dinosaurs. But as Ricqlès states, the bone histology in dinosaurs "impliquent peut-être la possession de l'homeothermie" ("may imply the presence of endothermy").[30] As that statement shows, Ricqlès puts forward a possibility, not a probability. In another work this author has expressed the opinion that "the big dinosaurs had a peculiar physiology by any standard, one which can hardly be regarded as 'typically reptilian' but must be better understood as something of its own."[31] That states the matter very well indeed. The dinosaurs certainly were not typical reptiles, but whether they were truly endothermic is far from established.

So the matter stands. Perhaps the dinosaurs were ectothermic, relying upon their environment for the efficiency of their bodily functions; perhaps they were endothermic, with a built-in temperature control. If the dinosaurs were ectothermic, then the story presented earlier in this chapter may give some idea of how they behaved on any given day of their lives. If, on the other hand, they were endothermic, their behavior would have been more independent of day and night temperatures, of sun and shade, of air and water. If such were the case, then the reader may imagine them as behaving in that particular respect more like some of the large land mammals, elephants for example, with which we are familiar.

There is much that we do not know about the dinosaurs, in spite of the abundant evidence of their fossil bones. There is much that perhaps we shall never know.

FIFTH MONTH

The Tread of the Giant

THE RAIN HAD BEEN INTENSE—part of a tropical storm that had passed through the jungle, bending trees before its fury and drenching the land with such a drumming downpour that at times the water swept through the forest and across the open places in blinding sheets. For Brontosaurus and the attendant herd it was a time of suspended animation; the big sauropods could only stand and endure the storm. They were cooled by streams of water coursing down their large flanks; indeed, they were cooled to a point at which they were no longer comfortable. The rain was washing away some of their body heat, so that they were just a bit sluggish. That was to their disadvantage.

If they had been near a swamp or a river, perhaps they might have retreated to the water when the storm broke. But they had been caught unawares and thus had to wait out the storm within the confines of the deep jungle. It had been a trying time, without food and without warmth.

But now the sun was out, and as usual the forest was steaming. The humid heat was bringing back some of the lost temperature of the dinosaurs, so they had again begun to wander and browse. Walking was not easy in the hours following the storm. The ground was muddy and slippery, and in places, too insecure for the giant reptiles, whose broad feet would each repeatedly bear down upon the earth with pressures of many tons. So their progress through the forest was slow and at times erratic.

Nevertheless they continued on their way. It was a case of putting down one gigantic foot after another, often with caution and deliberation. Almost always each foot was lifted and moved forward and was on the point of being put down again before the next foot was lifted. Three of the four feet thus maintained contact with the slippery ground, and never was any foot very far from the earth. A three-point support was important to these heavy reptiles on ground as treacherous as the terrain they were now traversing.

Although their progress was slow and at times halting, it had a pattern. First (A in the diagram) the right forefoot was put down, while the right hindfoot was being lifted and brought forward. As the right hindfoot was firmly planted against the earth (B) and muscular force was applied to push the body forward against that fulcrum, the right forefoot was lifted. The left hindfoot, now at the culmination of its backward thrust, was lifted (C), to be brought forward. While this was being accomplished, the left forefoot had been lifted (D) and brought forward.

If one were to diagram the sequence of footsteps on this page by showing the feet in contact with the ground, assuming that Brontosaurus was walking toward the right-hand side of the page the pattern would look like this:[32]

	Hind	Fore	Hind	Fore	Hind	Fore	Hind	Fore
Left	0	0	0	0		0	0	
Right		0	0		0	0	0	0
	A		B		C		D	

This illustrates how as each foot was lifted for a forward step, the body still was supported by three broad feet firmly planted on the ground. It was not a pattern for rapid movement, but it certainly ensured stability. Stability for Brontosaurus, with her thirty tons of flesh and bone to be simultaneously transported and protected, was a most important factor in her life.

Such was the pattern of footfalls, followed on a gigantic scale by Brontosaurus and her companions. Their march, however, involved more than the lifting and putting down of huge, elephantine feet. As Brontosaurus walked, her complex backbone, a dozen feet or more above the ground, swayed from side to side in a succession of sinuous movements. Between the shoulders and the hips the backbone bent according to the position of the feet. When the feet on one side were farthest apart, with the legs extended, the backbone was curved in that direction, and when those two feet were brought close together by the backward thrust of the forelimb and the forward swing of the hindlimb, the backbone curved away from this side toward the other. As the vertebrae of the back bent in a gentle curve in one direction, those of the long tail swung in the opposite direction. In that way the vertebral column from the shoulders to the tip of the tail was constantly bending in a series of alternating S curves, correlated with the movement of the limbs. It was a matter of muscular forces acting in sequence on the limbs and feet, the vertebrae and the ribs. In short, the whole body was involved.

When Brontosaurus walked, her long neck tended to swing back and forth, like the tail, in directions opposite to those of the swaying back. But the neck was not so intimately involved in the mechanics of walking as the back and tail; it was available for other purposes as well. As

Brontosaurus walked along she frequently swung her neck through differing arcs, in order to position her head variously as she viewed the scene in front of her. Also, her head was often being raised and lowered somewhat as it was swinging from side to side, so that movements in front of the shoulders were far from simple.

As may be imagined, tremendous forces were exerted against the ground by each foot as it was set down and as it helped to propel the body forward. Those forces—to be measured in tons—were carried from the body through the limbs and feet to the ground. For animals as huge as Brontosaurus, therefore, it was necessary for such forces to be exerted along the axes of the limbs, which is why the limbs of Brontosaurus were so massive and essentially straight. Such limbs were needed not only to support the body against the downward pull of gravity but also to withstand the stresses of locomotion.

The stresses of walking, immense though they were for a thirty-ton reptile, were not apparent to Brontosaurus. Behind her were many millions of years of evolutionary development through which, by ever-so-small stages, the adaptations to gigantic size had been developed. She had the strength of bone and muscle and ligament, in body and limb, to carry her great weight through the forests, across the glades, and into the marshes and rivers. And yet she had her limits, and she realized them in a dim, instinctive way.

That was why, when the sauropod herd, emerging from the jungle through which it had been walking, came to the edge of a small cliff, there was an immediate halt. But it was not quite immediate enough. A few of the big reptiles in the van of the herd felt the pressure from the sauropods behind them and attempted to brace themselves at the cliff's edge. The ground was muddy at the top of the cliff, so that try as they might, they could not stop themselves from slipping toward the edge, where they went over with a churning of ponderous limbs and a flailing of tails.

The cliff was not very high—in fact, no more than about twenty feet from top to base—but it was almost vertical, so that the few animals toppling over its edge fell to the bottom with earthshaking, echoing thuds. For the monstrous sauropods the fall was lethal; bones were broken, ligaments were torn, and huge blood vessels were ruptured. At the base of the cliff, which could easily have been negotiated by any of the small carnivorous dinosaurs—*Ornitholestes*, for example—the giants lay in the agonies of death.

Fortunately for Brontosaurus, she had been able to stop well back from the edge of the little cliff, and her great bulk, added to that of other large members of the herd, formed a barrier that stopped the sauropods behind them. So they stood, Brontosaurus and the forward members of the group, craning their long necks to peer over the cliff at the unfortunate victims below them. After a little time spent in indecisive milling about, Brontosaurus and the other sauropods finally turned at a right angle to follow the edge of the cliff, until it became a gradual declivity and finally died out completely. Once again on flat ground and surrounded by trees, the dinosaurs resumed their progressive browsing through the forest.

As the day drew to its close, the herd of sauropods marched on, by this time far removed from the scene where some of their number had met a tragic end. By now the incident at the little cliff had become a dim memory, if even that, in the primitive brains of the reptilian giants.

There is more to becoming a giant than merely growing large. The problem to be considered here is that of weight versus the strength of organic structures.

In the discussion of the second month it was pointed out that bulk increases by the cube of increasing linear dimensions, whereas surface area increases only by the square. The same considerations apply to the problem of the increase of weight and the correlative increase in size of supporting structures. Mass increases by the cube of increasing linear dimensions, but strength to support that mass increases only by the square of the cross sections of supporting bones. Consequently, there is a point at some place on the rising curve of increasing mass where the weight of the body is too great to be supported by its framework of bones. That point (never reached in the course of dinosaurian evolution) probably was being approached by *Brachiosaurus*, the 80-ton sauropod that enjoys the distinction of having been the largest animal ever to have lived on the land. The heroine of our story, Brontosaurus, was still considerably below that point; nonetheless she was a very large and heavy animal. (It must be remembered that some of the modern whales—the great blue whale, for example, more than 100 feet long and weighing 120 tons or more—are the greatest animals ever to have lived, far exceeding the dinosaurs in bulk. But the whales are supported by the water in which they live. A large whale that is washed ashore quickly dies because the skeleton is unable to support the weight of the huge mammal on land, and the pressure of the body on the lungs suffocates the stranded beast.)

The sauropod dinosaurs have massive bones forming the "appendicular skeleton": the limb girdles, the limbs, and the feet. The limb bones are large in cross section, and they are articulated with little flexure at the joints, to form long, columnar supports, rather similar to the limbs of modern elephants. The feet are broad and short and were provided in life with large, rather elastic pads to absorb the pressure of great weight and the shock when a foot with many tons above it came into contact with the ground. The evidence of the footprints clearly illustrates this.

In contrast to the appendicular skeleton, the "axial skeleton"—composed of the vertebrae and ribs, but particularly the vertebrae—is relatively light, yet shows remarkable adaptations for strength. The backbone of a sauropod dinosaur was in the neck region a long, powerful, flexible boom, as was shown in the discussion of the third month, while in the dorsal region between the shoulders and hips it was an exceedingly strong but flexible beam or rafter, from which in life hung many tons of weight: bones (the ribs), muscles, blood vessels, internal organs, and the contents of the digestive tract. The tail was also very strong.

As a means of combining strength with a drastic reduction of dead weight, the articular surfaces between the vertebrae of sauropod dinosaurs are expanded, but the bodies of the vertebrae are reduced in bulk by the development of large, cavernous regions in the sides of each vertebra. Many of the vertebrae in the column possess prominent ridges or arches between their articulating surfaces, paralleling in principle the flying buttresses of a medieval cathedral, which give strength to delicate, largely fenestrated walls. It is likely that the hollows in the vertebrae of the sauropods were filled with air sacs that were connected to the lungs. In addition to these adaptations for strength and lightness, each vertebra in sauropod dinosaurs has an extra pair of articular surfaces, called the *hyposphene* and *hypan-*

trum, which formed extraordinarily strong but articulating locking devices between successive vertebrae. Thus strength was achieved without the burden of extra bone.

The contrast in the sauropod dinosaurs between the unusually dense, massive appendicular skeleton and the almost airy-light axial skeleton led earlier authorities to postulate a "waterline" in these reptiles, running at the level of the backbone. The heavy limbs, it was argued, acted like divers' boots to give stability in deep water, while the light, strong backbone helped to make the dinosaur buoyant. That may be true. It does not necessarily follow, as some authorities formerly argued, that these great dinosaurs were confined to the water, that they could not come out on the land because the great weight of their bodies, unsupported by water, would crush the limbs. The footprints refute this argument; obviously the sauropod dinosaurs were quite active on the land and were adept walkers.[33]

It does seem probable, however, that the light backbone, with its possible attendant air sacs, gave buoyancy to these reptiles when they were in the water. The footprints at Bandera, Texas, indicating that the sauropods poled themselves along with the front feet, add what seems to be visible support to this supposition, showing that the sauropods were buoyant enough when in deep water. It is possible that additional air sacs in the pelvic region and at the base of the tail buoyed up that part of the body—hence the habit of poling with the front feet.

One function of the large, strong tail in the sauropod dinosaurs may have been to form a sort of prop, allowing these reptiles to rear up on their hindlegs, in spite of their massive bulk, for treetop feeding. The extraordinarily long spines on the vertebrae in the posterior portion of the back, in the pelvic region, and in the forward part of the tail might have been an adaptation for such a pose.[34] The spines would have allowed for the attachments of very strong back muscles.

So all in all the structure of the skeleton in the sauropod dinosaurs was admirably suited to give support and to withstand the tremendous stresses that developed as these giants walked across the land. They walked with confidence, yet it is quite possible that they walked with circumspection, too. They probably were instinctively aware of their great weight, just as today the elephant is aware of its weight. It seems, though, that there was nothing in the behavior of the sauropod dinosaurs to match the intelligence shown by the elephant in choosing a route. The elephant shows great sagacity as it maneuvers its bulk across the land; it knows fairly accurately what will support its weight and what will not. Although the sauropods did not have the elephant's intelligence, it is probable that the great dinosaurs were able to choose safe footing as they wandered far and wide. Their successful continuation through millions of years of geological history attests to this.

The description of fatal injuries to the sauropods that fell over the little cliff is based upon comparisons with modern elephants. These giant mammals, although far smaller than the large sauropods, are very susceptible to serious injuries from comparatively slight falls. A fall that would mean little to a small mammal or a mammal of medium size, such as an antelope, can be fatal to an elephant. The intrinsic strength of bone and ligament will not withstand the sudden shock of many tons coming down upon it with great force, even when that bone is heavy and dense. So the fatal fall of sauropods, even though over a low cliff, is here regarded as well within the realm of probability.

SIXTH MONTH

The Constant Wanderer

DAY HAD FOLLOWED day through the months since Brontosaurus had sought refuge from the savage attack by Allosaurus. During much of that time she had been constantly on the move, sharing her propensity for wandering with the other members of her herd. They moved together in an advancing wave of dark, gigantic bodies, staying close together for protection, especially for the protection of the younger and smaller members of the herd. They had to move in order to survive.

Then one morning the march of the sauropod herd came to a halt, because the big dinosaurs had reached a long, low seashore. In front of them, from a low eminence on which they were standing, they could see the sunlit waves stretching to the horizon. The low strand, where the land met the water, reached to the right and to the left.

It was an unfamiliar seashore confronting them, very

different from the shallow embayment where Brontosaurus had escaped from Allosaurus, and also different from that somewhat deeper embayment the herd had crossed, wading and swimming, in their journey from one bit of jungle to another. In their earlier encounter with the sea the dinosaurs had walked from a jungle onto a narrow, flat beach and into the clear water that separated them from another flat, narrow beach, fringed by another jungle. The transition from the forested land to the water and back to another forested land had been rapid, interrupted only by a short and easy crossing of sandy beaches. Here a new and strange meeting of land and ocean was disclosed to their reptilian eyes.

The sea was not nearby, even though it was plainly visible from where Brontosaurus and her fellows were standing; rather it was separated from them by a broad strip of undulating sand dunes forming a white, dry, hot barrier between the forest and the sea. It was a barrier that the dinosaurs had no desire to cross, because there was nothing beyond to beckon them. Only the breaking waves were to be seen beyond the dunes, and beyond the waves, the blue sky. There was no fringe of forest to interrupt the horizon, with its promise of plentiful leaves and fronds.

In spite of the rather erratic course the sauropod herd had traced during its months of wandering, the dinosaurs nonetheless had moved steadily toward the northwest. Months before, they had left an embayed shore where they looked to the southeast across the sea; now they had reached another shore, where the sea was to the north. They had crossed a broad land of low, rolling hills, covered by jungle and dotted with many little lakes and swamps, a land separating the southern sea that had been a part of their environment a half year earlier from the northern sea now in front of them.

This new sea, apparently so limitless, in front of and on both sides of the dinosaurs, actually was not a great ocean extending for thousands of miles from the land, but a long seaway reaching down from the north, only a few hundred miles wide from east to west. The shore, here oriented east and west beyond the broad strip of dunes, gradually and distantly curved north both toward the west and toward the east, to form the rounded end of the seaway. On the western side of the seaway was a comparatively narrow corridor of land stretching to the north, separating the seaway from a broad, western ocean. On the eastern side of the seaway the land extended as a great continent.

Of course these facts of geography could not be appreciated by the dinosaurs; to them the dunes and the sea in front of their position formed a broad barrier that they could not surmount. They had to turn either to the west or to the east in order to continue their wandering and feeding. And it so happened that they chose to go toward the east.

Having made the choice, they resumed their daily activities as before, walking through the forests and feeding from the treetops and from the ground. As they moved ever eastward, the land, though still abundantly clothed with green jungle, was less swampy than those miles they had traversed in their journey from the southern sea. For that reason Brontosaurus and her fellows fed almost exclusively on the trees and among the ferns on the ground, and had less recourse to water when the heat of the day dictated a diminution in their activity. Now, if they could not find a swamp or a lake, they rested quietly in the shade of the trees. Always on their left were the dunes, and beyond them the breaking waves.

Several days of this eastward progression had passed when one day the sauropod herd encountered another large herd of their own kind. The other herd had been walking and feeding in a westerly direction. The two groups came together in an abrupt confrontation within the depths of the forest.

But there was no animosity in the meeting. They met and mingled, and for some time they fed together. Finally, however, they parted ways. The herd that had been moving to the west continued in that general direction, while Brontosaurus and her herd went on toward the east. Neither herd traversed the pathways that had been followed before the encounter, because the forests they had left behind them were trampled and stripped of food. So Brontosaurus and her herd veered somewhat southeast after the two groups separated.

Actually the separation of the two herds was not a simple matter, because there were defections from and additions to each of the wandering groups of dinosaurs. Some of the younger members of the herd to which Brontosaurus belonged joined the westward-moving group of dinosaurs, while at the same time some individuals from that herd detached themselves to associate with Brontosaurus and her companions. The compositions of the two groups of dinosaurs accordingly were changed to some extent.

The changes were not great, nor were they dramatic, yet there was some significance to them, because the strangers in both herds were in the course of time to contribute "new blood" to the wandering groups with which they had attached themselves. Of course it was not new blood; it was the hereditary genes that were contributed by the newcomers to the two groups of dinosaurs. Both dinosaur herds benefited, because external genes were to be added to the gene pool of each herd, with the result that each group would be a little less inbred. Variety was to be added to the makeup of the two herds, and that was good.

This was only one of several such encounters, similar to encounters that had taken place in the past, as Brontosaurus and her herd moved along the shore of the northern sea. Several times in the future, as in the past, her herd would mix with another herd of sauropods, as it moved in one direction or another. Such meetings tied the herds together in one great network of wandering dinosaurs, reaching across the vast expanses of a great continent.

Brontosaurus could not know it, but the mixing and mingling that took place during the chance encounters between sauropod herds maintained her kinship, in a broad sense, with giant dinosaurs throughout the continents of an ancient tropical world. Perhaps in time she might mate with one of the newcomers to the herd; perhaps not. In any event, the newcomers formed links with the sauropod herds from which they had defected, so that between one herd and another the connections were maintained, to form overall a fluid population of the dinosaurian giants, extending for immense distances across luxuriant, green lands.

The description of slow forced marches of the sauropod herd, necessitated by the severe damage they inflicted upon the forests in which they browsed, is extrapolated from a general comparison with modern wild elephants. African elephants are notoriously destructive of the habitat in which they live, with the result that they are constantly moving from one area to another. In regions where the elephant population is dense, such as the Tsavo National Park in Kenya, damage to the bush can indeed be very serious. Iain and Oria Douglas-Hamilton, who for several years literally lived with wild elephants in Africa, and who have told of their experiences in a fascinating book, *Among the Elephants*, indicate that in extreme cases "35% of the trees were dead and more than half of these had obviously been killed by elephants."[35] Damage to the forest by elephants in that region today constitutes a pressing problem.

In that connection it should be said that Brontosaurus and her herd are depicted as wandering more widely than the elephants we know. Perhaps they did, and perhaps they did not. But the evocation of particular sauropod herds that wandered across some hundreds of miles of landscape may not be out of order.

It will be recalled that the story of the first months shows Brontosaurus and subsequently the herd of sauropods wading across embayments of a tropical sea. That story is based upon the rather spectacular sauropod footprints and trackways discovered near Glen Rose and Bandera, Texas. At the time the tracks were made, that portion of Texas was a shoreline along a southern ocean extending farther inland than the modern Gulf of Mexico. The sauropods evidently lived in the jungles bordering this ancient sea.

Their wandering to the northwest, to a northern sea, is based upon our present-day conception of late Jurassic and early Cretaceous paleogeography. In those ancient times a long and relatively narrow arm of the sea extended down from the north through what is now the Rocky Mountain region, into the general area of the Southwest. On the western side of this seaway was a narrow neck of highlands, separating the seaway from oceanic waters to the west. On the eastern side of the seaway were vast lowlands. One phase in the development of this southwardly extending seaway, the so-called Sundance Sea, was marked by an extensive dune area bordering the southern extremity of the narrow sea. The paleogeography of western North America, as interpreted above, has been made the setting for the present chapter.

Modern geologic information, based upon many lines of evidence in the fields of geophysics, paleomagnetism, geology, and paleontology, all seems to support most definitely the concept of plate tectonics and continental drift. That information indicates that during the middle and late Mesozoic Era, North America and Eurasia were broadly connected along a line that united what is now the eastern seaboard of North America with western and northwestern Europe, to form the ancient supercontinent of Laurasia. At the same time southern Europe was attached by a ligation between Spain and northern Africa to the ancient supercontinent of Gondwanaland, consisting of a vast landmass containing what are now Africa, South America, peninsular India, Antarctica, and Australia. Thus there were connections uniting the land masses of the Mesozoic world, so that land animals could move from one part of the globe to another.

Many of the land dwellers of that ancient world, including the dinosaurs, did move widely across the continents as they were then constituted. *Apatosaurus*, the genus to which Brontosaurus and her herd belonged, is now known from fossils found in North America, but perhaps it was more widely distributed than is indicated by extant evidence. Certainly that is true for *Camarasaurus* and *Brachiosaurus*, two of the sauropod types that Brontosaurus encountered in the chapter on the third month. *Camarasaurus* is known from western North America and from northern Europe as well. Excellent materials of *Brachiosaurus* have been found in western North America and in Tanzania, in eastern Africa. It seems obvious that this huge sauropod must have occupied a great distributional arc, stretching from Colorado and Utah, through what is now Europe, and down into Africa. The lack of specimens at the present time from European deposits may be one of those accidents of preservation, occurrence, and perhaps collecting that are common to paleontology.

As for dinosaurs contemporaneous with the giant sauropods, *Allosaurus*, characteristic of North America, probably lived in Africa. A closely related form, *Megalosaurus*, is found in Europe and eastern Africa. *Stegosaurus*, the plated dinosaur from North America, is closely matched by *Omosaurus* from Europe and by *Kentrosaurus* from Tanzania.

Thus the fossils show that some of the dinosaurs appearing in this story were very widely distributed, as a result of continental connections during the years in which they lived. Moreover, the presence of identical genera in North America, Europe, and Africa, as well as very closely related genera in the three continents, indicates that those dinosaurs probably did wander widely. That is not to say that any individual dinosaur walked from western North

America to northern Europe or beyond to eastern Africa, but rather that individuals did move about over considerable distances, meeting other individuals which in turn wandered and met other dinosaurs that carried the contacts on—and so on, across thousands of miles. In this manner there was a chain of individual contacts that frequently carried the gene pool across wide continental regions.

SEVENTH MONTH

The World Outside

BRONTOSAURUS STOOD in the van of the herd, looking out across a landscape that stretched for many miles in front of her. Beyond a small clearing in the immediate foreground was a vista of tropical forests and swamps. The herd, which some time previously had veered away from the sandy shore of the northern sea, was now marching back to the south, toward the crystalline ocean that they had left several months ago. Their journey, however, was being made somewhat to the east of the route they had taken when they had wandered north. It would seem that all in all theirs was an aimless journey—first to the north and west, then briefly toward the east, and then back in a southerly direction—and perhaps it was. But why should it be otherwise? They had no goal to achieve; their purpose was to go where there was food, and such a purpose, if so it should be called, led them through the jungle in a straggling and seemingly never-ending expedition.

There may have been subtle factors that determined the change in direction back to a southerly course. Perhaps the great reptiles felt the small differences in temperature between the country they were now traversing and that land from which they had come. Perhaps there was some sort of instinctive memory calling them back to their more familiar home, just as in our modern world many birds are beckoned toward the north or toward the south when the time of migration arrives.

However that may be, it was here that the sauropod herd paused, looking south, toward the equator, which in those distant days was not far removed from the southern sea along whose shores (in present-day Texas) the dinosaurs had spent so much time. And as the herd paused, Brontosaurus stood in the lead, scanning the scene that confronted her.

It was a forest scene of the kind she had looked at so frequently during her long life. It was a scene dominated by the greens of the jungle, with here and there splashes of other colors: the varied florescences of cycads, the russet colors of some of the tree-fern fronds, and the arresting reddish tips of giant horsetails. Above the expanse of the green jungle was a brilliant blue sky where towering thunderheads rolled—white and gray.

Did Brontosaurus see all of those colors with her large, reptilian eyes? They were excellent eyes, capable of discerning objects far and near in detail. Did they also comprehend the full range of color in the forest ahead? Was she able to see what was before her in the wide expression of its grandeur? Whether or not hers was a world of color, how did her small, reptilian brain interpret the scene?

It seems safe to assume that there was no aesthetic appreciation in her survey of the jungle toward which she and the sauropod herd were making their way. The forms and perhaps the colors were to her signals for browsing or for potential danger. They were the substance of the external environment to which she had to adjust the instantaneous, the hourly, and the daily activities of her huge frame. It was the outer world revealed to her through her senses—through her eyes, her ears, and her flaring nostrils. It was also a world of touch and taste, but the latter sensations were to be obtained only through intimate contact with the things around her. At that moment the outer world with which she was concerned was a distant world, a world to be perceived, to a considerable degree, visually.

That was why she had paused before continuing on her journey, and why the herd, of which she was a leader, had paused behind her. It was a time to reconnoiter the immediate foreground that was to be traversed by the wandering dinosaurs.

Her survey of the scene ahead was not, however, limited to a visual inspection, because she listened as well. She had excellent ears to supplement her excellent eyes, and she constantly put those ears to good use. The large eardrums were each more than two inches in diameter and largely exposed behind the articulation of the lower jaw on the skull. They vibrated when sound waves impinged upon them and, by means of a long, slender bone attached to the drum on each side, the stapes, transmitted the vibrations to the inner ear. It was in that manner that sounds, so very important to Brontosaurus, were brought from the outside world into the innermost recesses of her arched skull.

She listened intently as she looked across the landscape in front of her, and she heard many things. There were the chirpings and the countless other noises made by insects, carried to her ears on the hot, drowsy air. There were the croakings of frogs in the marshes, and even occasional bellows from crocodiles in these same marshes. There were the hoarse calls of primitive, long-tailed birds, the early precursors of the feathered host that was to be so predominant in forests of later ages. And there were the noises made by other dinosaurs, for all kinds of dinosaurs probably communicated audibly. However, amid all of these sounds she could detect nothing that predicated danger.

Finally, she tested the air with her nostrils. Whether they gave her much information it is hard to say, but at least they did serve to supplement whatever knowledge of the world had already been gathered by her eyes and her ears. The scents that were borne to her all seemed to bring a promise of good things ahead; they were the scents of the forest—of green vegetation, lush and ready for browsing.

The messages received by eyes and ears and nostrils were positive, and as such they decided the immediate course of action on the part of Brontosaurus. She stepped forward, her large padded feet clumping down on the solid earth, her tough, leathery skin wrinkling and unwrinkling as she swung her massive limbs back and forth, her tail dragging on the ground, leaving a wake of crushed vegetation. As she walked she constantly swung her long neck from side to side, in order to view the route ahead from successively different angles.

Her actions were imitated by the other members of the herd, all walking toward the fresh-smelling woodland ahead, all swinging their heads from side to side as they walked. As was their custom, they marched with the larger members of the group on the outside of the gathering, to protect with their huge flanks the smaller and weaker dinosaurs occupying the center of the marching herd.

The herd entered a clearing, making for the jungle beyond and its promise of rich browsing. The sun shone brightly, and in the clearing, exposed as they were, the moving group of dinosaurs formed a dense procession of great forms, towering above the low vegetation. To many it would have been an impressive sight, indeed an awe-inspiring vision. But the eyes that were watching the ponderous march were avid. They belonged to four hungry allosaurs, great predators of the same kind as Allosaurus, who had attacked Brontosaurus in the marine embayment so many months before. These allosaurs watched the sauropod herd from the protection of the forest margin at the edge of the clearing with a mixture of excitement and indecision. They wished to attack, but they were somewhat daunted by the combined mass of the marching sauropods, especially by the size of the flanking marchers. But the driving urge of hunger overcame any such discretion as might have been summoned by their primitive brains, and they attacked.

They charged toward the left flank of the sauropod parade, emerging from the sheltering forest with savage speed. Their sudden appearance disconcerted the usually placid sauropods, and for a few brief moments there was consternation and confusion within the ranks of the giant dinosaurs. They began to rush toward the margin of the forest in a rather disorganized, noisy mass, plunging through the low plants of the clearing with a thunder made by hundreds of tons of giant fugitives, pounding the earth with their broad feet. Added to the noise of their trampling flight were the cries they made as they ran—grunts and bellows of fright.

This brief lapse into disorganization within the herd

worked to the advantage of the four predators. They dashed through a gap that had opened up between two of the flanking sauropods, to attack one of the younger reptiles in the middle of the herd, a successful—although hardly premeditated—maneuver on the part of the allosaurs. With their combined force and savagery they pulled their victim down to his doom. The attack, although it would ultimately be fatal to the sauropod of their choice, was of no lasting consequence to the other members of the herd. Diverted as they were by their concentration on their prey, the predators paid no attention to the other giant dinosaurs, so that the fleeing reptiles made their escape. There was no urge among the retreating sauropods to defend the unfortunate victim. Sentiment is not a reptilian trait; the concern of the herd members was, individually and collectively, to save themselves. So they lumbered on, entering the forest in broken ranks, soon to be reorganized. Their flight continued for some little distance before they felt secure enough to slow down and finally to stop.

In the meantime the four hunters were having problems with their prey. The sauropod they had attacked was not fully grown, but nonetheless he was big and strong, and it took a great deal of effort to subdue him. Two of the allosaurs had seized him by the neck; it was slender enough that they could grasp it with their great, gaping jaws. They held on like grim death, which indeed it was, while their companions slashed at the victim's flanks and limbs. As a result of their combined efforts, the sauropod finally went down with a resounding crash.

But the struggle still was not over. One of the allosaurs that had obtained a hold on the sauropod just behind his head had finally bitten through the spinal nerve cord, thereby severing communication from the brain to the body. The neck of the sauropod went limp and the forelegs lost their mobility. In a sense the sauropod was dead—at least the forward part of the big reptile was dead. The hindlimbs and the tail, however, were far from dead; they thrashed around with tremendous force, throwing off the attackers time and again by the violence of their movements. Even though the brain had been cut off from the body, the tail and hindlimbs retained their autonomy by virtue of the great sacral plexus, a massive, complicated enlargement of the spinal nerve in the region of the hips for the control of the huge hindlimbs and the tail. The sacral plexus, even though much larger than the brain, was not a second brain, but rather an accessory control that acted somewhat independently of the brain. Thus the back portion of the dying sauropod continued to function, in spite of the lifelessness of its head and neck. So the struggle continued, filling the clearing with action and noise, trampling and crushing the vegetation, and soaking the ground with blood.

It was a long time before the affair was ended; not until after the sauropod was cruelly torn and partially dismembered did he finally expire. From that point on the clearing was the scene of voracious feeding. The four predators tore out great chunks of flesh to devour them whole. Sometimes, when one of the hunters had ripped off an especially large piece of the victim, the fragment would also be seized by another allosaur, and the two predators would pull and twist the flesh until it was torn into two pieces, then gulp it down. In that manner the gory feast continued. The four allosaurs, working together on a less than full-grown sauropod, had been able to accomplish in concert what Allosaurus had failed to do in his solitary attack on Brontosaurus, when she was crossing the embayment of the sea.

Meanwhile the herd, with Brontosaurus in the lead, continued on its way. By now the attack was a thing of the past, and the sauropods began to browse, as had been their intention when they first entered the clearing on their way toward the treetop pastures of the jungle. These pastures were good, and they browsed, off and on, through the day, entering a convenient swamp in midafternoon for their usual cooling-off period.

Thus the day ended, a day punctuated by tragedy and fright, yet overall a day of no great consequence to the herd. One of their number was missing, but it was a matter of little significance to the total group. Such unfortunate depletions might occur from time to time, but in the end the herd maintained its size and composition. There were subtractions from the mass, but there were additions, so life went on, with its usual monotony and its occasional incidents.

It is very difficult, of course, to know what animals other than ourselves see, or how they interpret what they see. From varied experiments it would appear that some modern reptiles have color vision (as do some modern mammals), while perhaps other reptiles in this modern world are color-blind (as are many other mammals). As Angus Bellairs said, "Many lizards such as agamids and lacertids can certainly distinguish between different colours, as one might expect from the importance of colouration as a secondary sexual character. If lizards (*Lacerta agilis*) are offered as alternatives tasty mealworms and unpleasant salty ones on discs of different coloured paper they can learn to distinguish red, orange, yellow, yellowish-green, blue and violet from each other and from various shades of grey. . . . Giant tortoises can be trained to distinguish between orange, blue and green, and certain terrapins appear to be particularly sensitive to colours of long wave-lengths towards the red end of the spectrum; they may even be able to perceive infra-red radiations. Crocodiles and snakes, on the other hand, are probably colour-blind." [36]

How, then, are we to determine the presence or absence of color vision in the dinosaurs? Did they have some color vision like the lizards and turtles, or were they color-blind, as seems to be the case with their closest reptilian relatives, the crocodiles? One might argue in favor of the latter alternative, except for the fact that birds, which according to modern evidence are directly descended from theropod dinosaurs and are in a sense as closely related to the dinosaurs as crocodiles, are richly endowed with color vision. Did color vision evolve in the birds after their separation from their dinosaur ancestors, or was that attribute developed in common among dinosaurs and birds? This is a field for much speculation.

It is probable that the dinosaurs had excellent vision, whether it was in color or monochromatic. Bellairs wrote that "most diurnal reptiles which lead active lives above the ground probably rely more on their eyes than on any other sense organs. . . . They respond, like many other keen-sighted animals, much more readily to moving objects than to stationary ones, though it is possible that here the brain rather than the eye is involved." [37] It is reasonable to suppose that Brontosaurus saw things in detail, near and far.

The crocodiles possess a special layer, the *tapetum*,

within the eye that reflects light back through the visual cells, thereby allowing for the utilization of a greater proportion of light entering the eye. These reptiles have excellent night vision. Were the dinosaurs similarly adapted? Again, we can only speculate.

It seems probable that like many modern reptiles (and birds as well), the dinosaurs possessed a narrow field of binocular vision. (For example, the field of binocular vision is 25 degrees or less in the crocodiles.) [38] It is not unreasonable to think that dinosaurs like Brontosaurus were constantly moving their heads, the better to see things with one eye or the other, or with both eyes in conjunction. And what they saw frequently determined their responses.

That last consideration would apply to the allosaurs that are described as making a concerted attack on the sauropod herd. Wrote Wilfred Neill, "It must be emphasized that the alligator never evaluates or responds to the totality of any situation, but merely reacts automatically to some particular stimulus that is provided by that situation." [39] The situation confronting the allosaurs was that of a moving sauropod herd. The predators probably were driven by a consuming hunger, the sauropods were moving, and this movement of the big dinosaurs may have triggered a response on the part of the allosaurs. They attacked without any evaluation of the situation or without any plan. As luck would have it, they were successful.

Some dinosaur skulls have been found with the stapes bone in place. This bone, together with a cartilaginous appendage, the *extrastapes*, transmits vibrations from the eardrum, or *tympanum*, to the inner ear in modern reptiles, performing the same function that is accomplished by three bones in the mammalian ear; the *malleus*, *incus*, and *stapes*. In spite of the comparative simplicity of the reptilian mechanism it is quite adequate, and many reptiles have an excellent auditory sense. Such would seem to have been the case in the dinosaurs. The stapes in those reptiles is long and slender, bridging the considerable gap between its connection with the tympanum and the opening in the braincase, the *fenestra ovalis*, within which its inner end was seated. It was admirably formed for the transmission of vibrations.

It is likely that in most dinosaurs the eardrum was large, and probably sensitive. As is general among reptiles, the drum was probably largely exposed; there was no complicated outer ear, or *pinna*, to protect it, as is characteristic of mammals.

The dinosaurian brain was relatively small and primitive. The brain in Brontosaurus, as determined from casts, probably was about 0.001 percent of the body weight. Thus if Brontosaurus had a live weight of about 30 tons, or approximately 30 thousand kilograms or 30 million grams, the brain would have weighed perhaps 300 grams. She may be contrasted with the Nile monitor, a very large lizard, with a brain-to-body percentage of 0.03, or 2.4 grams to 7,500 grams; the green lizard, a small reptile, with a percentage of 0.4, or .13 grams to 32 grams; and the cat, with a percentage of 1.0, or 30 grams to 3,000 grams. In humans the percentage is on the order of 2.1, or 1,400 grams to 65,000 grams.[40]

Size alone is important, but it is not the only factor to be considered. The modern reptilian brain is primitive, and so was the brain of the dinosaur.

EIGHTH MONTH

The Return

THE SOUTHWARD MOVEMENT of the herd continued, and as the days passed into weeks during the migration toward their old homeland, the air became ever so slightly warmer and the jungles ever so slightly more humid. The differences were not great, but in a cumulative way they could be felt by the band of giants. The warmer, balmier air spoke to them of their home territory, so that beyond the necessity for feeding this was a time of constant marching—a time to go home.

Yet there was another urge, stronger than the call of warm temperatures, that beckoned them to the south. It was the urge to reproduce, to perpetuate their kind, and for these particular dinosaurs the locale for founding a new generation was along the shores of the sunlit southern sea. Hence the rather purposeful aspect (if so it may be called) of their return to the land of their origins.

The powerful force that pulled them to the southeast did not act suddenly or capriciously. It began, perhaps faintly, as they began to leave the northern sea, and it became increasingly stronger and more constant as their journey continued. Theirs was to be a long journey, and they were to live with this force day and night. It pushed them on despite the daily lingering for feeding, and despite distractions along the way.

Some of these distractions were other dinosaurs, of their own kind and of other kinds. As for their own kind, there were the peaceful confrontations such as they had already experienced more than once during these past few months. They would on occasion encounter another wandering herd of brontosaurs, and there would be the mingling, the joining, and the breaking apart again to which they were accustomed, with the usual loss of some of their members to the group of strangers and the usual recruitment of several individuals from the other herd to their own assemblage.

As for other kinds of dinosaurs, it was usually a case of a meeting and then a passing on. Brontosaurus and her fellows would walk past or through a group of camarasaurs without giving these smaller sauropods much attention. They would view a *Diplodocus* or two in the distance, usually half submerged in the still waters of a swamp. They would make way for a huge *Brachiosaurus*, or perhaps for several of these gigantic sauropods—not through any particular sense of fear, but rather through a sense of inferiority. The plated, spiked stegosaurs were passed by. The inoffensive camptosaurs were largely ignored. And the little, agile carnivores, such as *Ornitholestes*, scampered from the path of the advancing sauropods. The frequent confrontations did not materially alter the progress of Brontosaurus and her herd in their chosen direction.

Two or three times, however, during the course of their southern journey they were subjected to attacks from dreaded allosaurs. The attacks were sporadic, and their effects were various. On one occasion the herd escaped without any injuries. On another, a small brontosaur that unfortunately lingered a few paces behind the herd paid the price for its inattention. It was overwhelmed and devoured. Yet often enough the allosaurs lurked on the edge of the herd without pressing an attack; Brontosaurus and her companions maintained themselves in such a compact mass of gigantic bodies that the predators were prevented, in their instinctive manner, from attempting a charge against the huge moving wall of sauropods.

Also, during the time when the brontosaurs were moving to the southeast, one of their number dropped out. He was an old reptile, too feeble to stay with the herd. His joints were arthritic and swollen; his movements, slow and labored. Gradually the ancient sauropod fell behind, to toil slowly through the forest, feeding from those trees to which he could painfully stretch his somewhat inflexible neck. In that manner he continued on his solitary path, each day farther and farther removed from the herd. The poor old veteran eventually became quite lost and probably suffered the common fate of the old and the infirm in the world of nature—which was to be attacked and killed by a roving predator or group of predators. The herd continued on its way, diminished by one, yet scarcely aware of the loss.

As the herd moved ever farther to the south and east a general feeling of unrest and tension slowly developed among the more mature members of the group. Brontosaurus shared this feeling with her peers, and as the days followed one after another the feeling grew ever stronger within her. And although she became more unsettled as the days progressed, she showed few outward signs of her inner turmoil.

Such was not the case among the large male sauropods. Their inner feelings of unrest were expressed outwardly by changes in their behavior. They became more aggressive toward each other and toward other members of the herd, whatever their age or sex. They pushed and shoved. Frequently two males would face each other, to nod their heads up and down or weave them back and forth through the considerable arcs afforded by their fifteen-foot-long necks. At times they would entwine their necks as they pushed against each other. All of this was very noisy business. They bellowed and roared in a manner designed to frighten, if not each other, at least some of the lesser members of the herd. In short, the males were trying to establish positions of dominance—the dominance of number one over number two, of two over three, and on down the line.

The bellicose exchanges among the otherwise gentle giants were indications that the season for mating was near, and as the crucial time drew nearer the demonstrations of the big male sauropods became louder and more aggressive. In fact, there were times when the pushing and shoving, the demonstrating and the bellowing, became so obsessive that these males would forget to browse and to march along with the quieter members of the herd. They would go through their patterned behavior while the other sauropods continued to walk and to feed. Then the fractious males would hurry to catch up with their fellows.

Finally the rivalry among the dominant males reached its climax. The competition among them could go no further, because they had established among themselves their own priorities for desired females. From now on, having established something of an uneasy truce within their self-imposed hierarchy, they turned their attention more and more to the chosen females of the herd. They picked their mates according to the positions they had established by their prolonged quarreling, with the most dominant males choosing the most females, the less dominant males necessarily having to be content with fewer mates, or even a single one, while many of the younger males were consigned to bachelorhood. Not that the pairing or the building of small harems of sorts lasted for any great length of time; the arrangements were temporary. But while they lasted they were effective.

All of this activity among the males stimulated the females, making them receptive to advances from the contesting warriors when at last the antagonists focused their attention on the prospective mates.

That is how Brontosaurus found herself one day, in company with two or three of her female companions, under the close care and supervision of the largest male in the herd. He pushed them along, and kept them slightly apart from the other dinosaurs, and seldom was he more than a few paces from any one of his little group. For a day or so they traveled together, close by but not intimate. Then, as their desires reached a peak, copulation took place. The big male mounted Brontosaurus, and in turn the other females that he considered to be his by rights—and that was that. As soon as the matings had been accomplished, the desires among these particular sauropods for closeness, for mutual companionship, quickly died down, finally to vanish. And so it was with the other males and females of the herd. The matings extended through several consecutive days; then they were finished. Once again the herd became a single coordinated body, rather than a union of little subgroups, as it had been during the days of mating. Once again the dinosaurs marched together toward the south.

Through the time of courtship rivalry and of mating, the herd had been moving in its chosen direction, but perhaps not in so determined a manner as when they first turned away from the northern sea. Now the march was again continued in a purposeful manner; they were returning home and they were almost there.

The idea that a herd of sauropod dinosaurs might wander in one direction, and then turn around to retrace its steps back to the place from which it came, is of course conjectural. Yet it is not outside the realm of possibility. Certain modern reptiles, notably the giant sea turtles, make unbelievable migrations across the ocean. Of course the Mesozoic landscape was a locale quite different from the South Atlantic of today, across which marine turtles make such long journeys. But mightn't large, active, herbivorous dinosaurs, such as the sauropods, have made long journeys on the land?

It is here assumed that they did, and that in the course of their travels to the north they encountered, perhaps in New Mexico, the Sundance Sea—an unexpected obstacle. Then it is supposed that they experienced an urge to return to their homeland near the southern sea, whose shores crossed what is now eastern Texas from northeast to southwest. The migration would have been on the order of several hundred miles in each direction, but if the great sauropods moved along at about five miles per day they certainly could have made the trip to the north and back to the south within the better part of the year that this story encompasses. They had to keep moving to feed, but if they kept wandering in a generally consistent direction, they might find themselves in a strange country, facing an unknown sea.

The idea of the urge to return has its basis in part in what we know about the migrations of the sea turtles. When nesting time comes, these remarkable reptiles cross thousands of miles of ocean to land on a particular strip of beach, perhaps no more than a mile or two long. How they navigate so precisely is a mystery as yet unsolved; Dr. Archie Carr, a distinguished herpetologist at the University of Florida, has been looking for the solution for many years. The urge to go back to the land of origin is based also in part on the terrestrial seasonal migrations of various large, hoofed mammals—of large antelopes in Africa and formerly of the bison in North America. Such journeys cover many hundreds of miles in a season.

The account of an old dinosaur falling behind because it was weakened by senility and arthritis is quite in keeping with paleontological evidence. The evidence of arthritis is commonly found among fossil vertebrates, including dinosaurs. Arthritis is an ancient disorder. It has been around for a long time, and there is every reason to think that it will be around for a long time into the future.[41]

If the story of wide wandering by the herd of brontosaurs is conjectural, the reconstruction of courtship patterns and of mating is perhaps even more in the realm of conjecture. How can we know how these dinosaurs behaved during this crucial annual period? We can only look at modern vertebrates and try to visualize what things might have been like during the Mesozoic Era. The reconstruction of behavior patterns presented on the preceding pages may not be correct, but it may not be wildly erroneous, either.

Let us turn once again to the closest modern reptilian relatives of the dinosaurs, the crocodilians. What we know about the mating habits of these reptiles is confined almost exclusively to the American alligator and the Nile crocodile. Here some of the evidence is not completely reconcilable. For example, according to Wilfred T. Neill in his excellent book *The Last of the Ruling Reptiles*, "There is not a shred of actual evidence [among American alligators] to support the oft-repeated statement that 'rival males' battle over a female at this season."[42] Yet in opposition to that statement, there is the following observation by Hugh Cott in his monographic study of the Nile crocodile: "However the fact is well established that inter-male rivalry frequently finds expression in combat. Pitman states that crocodiles are notorious for fighting among themselves, and that where they are abundant they indulge in terrific contests, which often terminate fatally. . . . Some of the inter-male fighting at breeding time may be for the possession of mates. According to Deraniyagala (1939) males of *C[rocodylus] porosus* [the salt-water crocodile] are said to fight each other for the females; Clarke (1891) refers to fierce battles among males of *A[lligator] mississippiensis* in the breeding season. . . ."[43]

As for courtship, certain patterns of behavior have been observed among the crocodilians. These consist especially of the male stroking the female with his forelimbs. According to Neill, "During the courtship period the male remains with the female until she is stimulated to permit copulation. It usually takes at least three days, and sometimes up to seventeen days. . . ."[44] Courtship patterns in other reptiles, especially lizards and snakes, may be rather elaborate. Copulation in the crocodiles is brief.

If dinosaurian behavior patterns paralleled to any degree what is seen in some of the large mammals, there is reason to think that inter-male rivalry may have been lively, if not intense, and that courtship patterns may have been developed with some degree of elaboration.

With such considerations in mind, it seems logical that the behavior of the large male and female sauropods was as reconstructed in our story. Certainly breeding behavior in these dinosaurs was effective; they populated the earth in abundance through a span of many millions of years.

NINTH MONTH

The Nest

THEN ONE DAY, there it was—the southern sea, crystalline in the early morning light, with the slanting rays of the sun reflected from innumerable thousands of wave crests that glittered to the horizon in ever-changing ranks. The herd had completed its circuit from southeast to northwest and back again, from the strand of the southern sea to the uninviting dune-bordered limit of the northern sea and then back to more familiar shores. The dinosaurs had for this year come to the end of their long elliptical migration between seas and through jungles. Now they were home, if such creatures can be said to have had a home, to wander along the edge of their own particular ocean where they might feed in surroundings hitherto spared the destructive effects of their passage.

For Brontosaurus and those other females who had mated there was a purpose, beyond mere feeding, in this return to the southern ocean. It was for them a time of nesting, a time when the future of their kind took precedence over the mere necessities of sustenance. They knew instinctively that it was a time for doing certain things in a fixed sequence, and instinct told them how those things were to be done. While the males, some of the females, and the younger members of the herd went about their feeding in the usual way, those particular females, with Brontosaurus in the lead, diverged along a path of their own.

They were searching for a proper place for making nests and laying eggs, and it had to be just right. Brontosaurus and her sauropod sisters tramped along through the jungle and past some swamps, looking for the nesting place. Two or three spots were examined and rejected; instinctively they knew that those places were not what they needed. Either they were too sunny, or too shady, or perhaps the ground was not of a consistency to suit them. So they went on.

Finally, after searching for some hours, the group of females found the right place, and then their labors began. It was a sandy area, back from the seashore, and not far from the fringing forest. Located between forest and sea, the area was one of partial sun and shade where isolated trees or clumps of trees and little thickets of low vegetation dotted the scene, so that through much of its extent the sunlight was filtered before it reached the ground. The sandy surface was in many places warm, but not too hot. It was an ideal place for sauropod nests.

Brontosaurus picked a place to her liking, between two large tree-ferns and bounded on one side by several spreading cycads. Here she circled about several times and then started to dig. As she worked, other females around her began to make their nests, while beyond them more sauropods were busy digging. Each female had chosen her site carefully, paying close attention to the balance of sun and shade at the place where she was digging, and giving particular notice to the distance she might be from her neighbors. The laboring sauropods spaced themselves across the nesting area, and in general they maintained the distances from one another that their instincts told them were proper.

Brontosaurus scooped a large, craterlike depression in the sand with her massive, clawed hindfeet. She stood firmly at her post, pushing back and outwardly with one foot, and then with the other, digging deeply into the sand and forming a ring-shaped mound of material around the edge of her excavation. After having dug down some distance in one position she rotated her body about 90 degrees, to continue the excavation in another direction. So she proceeded, digging and turning, all the time making the nest not only larger and deeper but also increasingly circular. The work went on until she had completed the nest to her satisfaction, which meant that it was in the end a large crater, perhaps fifteen feet or so in diameter and several feet deep at the center. Around the edge of the crater was a rim of sand that had been removed from the excavation.

It was a large nest, and for good reason; Brontosaurus was a very large dinosaur. Some of the smaller females made smaller nests. For the most part the size of each nest was roughly in proportion to the size of the dinosaur making the nest. Furthermore, the number of eggs to be deposited in each nest was dependent to a considerable degree upon the size of the dinosaur laying the eggs.

But Brontosaurus had not as yet begun to lay her clutch

of eggs; she was still perfecting the shape of the nest. This required many finishing touches, so that the nest would be well formed: when completed, it would be almost perfectly circular in outline. Also, she had to spend considerable time fussing with the inner slopes of the nest—to remove sand that inevitably slumped in during her nest-building activities to mar somewhat the desired uniformity of the crater. Her instincts, which so strongly controlled the nest building, told her what to do, and she carried out those instructions with meticulous detail.

Of course she was not alone in the careful task of perfecting a nest; the sauropods all around her were busy with the final details of nest construction. Nearest to her was another large female making a nest almost as large as her own, of which the nearest edge was only a few feet from the nest of Brontosaurus. And only slightly farther away, on the opposite side, two much smaller nests were being excavated by small females. Perhaps these were the first nests of their careers.

By this time the nesting area was rather profusely dotted with the craters that had been dug into the soft sand—some of them large, others smaller. Indeed, the sandy surface at this stage resembled on a small scale the face of the moon, pockmarked with varying circular craters, each sharply delineated by its raised rim. Yet in spite of the moonlike appearance of the area, one very definite feature characterized the arrangement of the nests: none of them overlapped. They were spaced in such a way that an area of undisturbed ground, however small, surrounded each nest. In a sense, each nest had its own territory.

The spacing of the nests was not, however, arbitrary or in any manner symmetrical. Each had been sited with relation to the vegetation of the area; each female, like

Brontosaurus, had placed her nest so that it would receive both sun and shade, so that it would not become too hot beneath the rays of the midday sun, or too cool in the projected shade from bushes and trees. The arrangement of the nests varied from fairly close craters, where there were clumps of trees or thickets of low vegetation, to widely spaced nests where shade-producing vegetation likewise was scattered. Some portions of the nesting area, devoid of vegetation and thus subject to the intense light of the sun, were devoid of nests.

And now the time had come for Brontosaurus to lay her eggs. She had finished her nest. In that respect she was somewhat ahead of most of her fellows, even though she had dug a very large nest, perhaps because she was a female of experience. She had done it before. She prepared herself for egg laying while some in her vicinity were similarly occupied, but while others were still busy digging.

Backing up to the big crater that she had excavated, she placed her hindfeet at the edge of the nest, swinging her long neck in a lateral arc in order to peer back for a look, to see that the positioning of her feet was right: close enough to the nest that the eggs would drop properly into the crater, yet not so close that her great weight would cause the crater to cave in. It was a delicate operation requiring precise control of her thirty-ton body and especially of her massive hindfeet, but she managed it nicely. She managed in part by bending her hindlegs so that her body was lowered, bringing the vent immediately above the nest. Then the process of egg laying began.

The eggs dropped down, one by one and in rather rapid succession. They did not have far to fall, and the impact of each egg as it dropped into the nest was cushioned by the soft sand on which it landed. Each egg was round, somewhat larger than a large grapefruit and somewhat smaller than a bowling ball. The strong, white shell of each egg had a rough, pebbly surface, giving it a texture somewhat like that of very coarse sandpaper but with the individual eminences smoothly rounded. By virtue of the rough shell there was a certain amount of friction between each egg and the sand as it dropped into place, so that usually it did not roll into contact with the egg that had preceded it. Brontosaurus continually moved her vent during the deposit of the first eggs by gently swaying her pelvic region back and forth and forward and back, with the result that these eggs formed a rough circle near the bottom of the crater. Perhaps a half-dozen eggs were so positioned to form the circle.

Then she kicked some sand over these eggs with a hindfoot and proceeded to drop more eggs outside of and slightly above the first eggs, in a larger circle of a dozen or so eggs. More sand was kicked in. Then still a third circle of about eighteen eggs was deposited, one by one in quick succession. Again more sand, and finally a last, large circle of about two dozen eggs was dropped into place, this last operation requiring a considerable amount of shifting her body, and even moving her feet, to facilitate the process.

In this manner four circles of eggs, about sixty in all, had been placed in the nest, each circle outside, and slightly higher than, the circle it succeeded, each circle conforming to the sloping inner sides of the crater, and each circle separated from its contiguous circle by sand that had been shoveled back into the nest during the complicated process of egg laying. Then, with all of the eggs in place, Brontosaurus proceeded to fill the crater with the remainder of the sand on the crater rim. She did this methodically and carefully, using both hind feet to push the sand carefully onto the nest until its top was flush with the ground surrounding it. After this, she stepped forward and swung her tail back and forth like a broom, until all traces of digging had been removed—until it was scarcely possible to see where the nest had been.

After that Brontosaurus abandoned the nest that she had so carefully built and the five dozen eggs she had so carefully laid and concealed, to search out the herd that had gone on feeding through the forest. She was one of the first females to complete nesting, so it was not long until she had found and rejoined the herd. As she again began to crop leaves from the trees around her, other females kept arriving to join the browsing sauropods. Some of the arriving females were large, experienced individuals that had completed the task of building their nests, laying the eggs, and covering and concealing the nests in a forthright and efficient manner. Other females were younger and smaller, and since they were less experienced, they had taken more time for those important duties, even though the nests they had built were relatively small, and the eggs placed in those nests were relatively few in number.

Eventually all of the herd was reunited, and the entire body of dinosaurs drifted through the forest, browsing among the treetops. The nesting site was left behind, seemingly forgotten, a level stretch of sand beneath the scattered trees and bushes, marred here and there by gigantic footprints.

The delay of a month or more between the mating of Brontosaurus and her nesting and egg laying is extrapolated from what we know about modern reptiles. In the American alligator mating takes place during the early part of April, yet the building of the nest does not begin until the first part of June, making a two-month delay between coition and egg laying. This is not surprising, since it is known that among many modern reptiles there can be prolonged storage of sperm within the genital tract of the female. It is particularly true among turtles, lizards, and snakes—sometimes, in the case of certain turtles and snakes, extending over a period as long as four or even six years. Whether the interval between coition and nesting in the crocodilians is a matter of sperm storage and delayed fertilization has not as yet been determined. Nonetheless, the delay between mating and laying eggs is a fact; it is therefore reasonable to suppose that such a phenomenon might have been common among the dinosaurs. This is the basis for having the mating of Brontosaurus take place perhaps a month before she and the other members of the herd reached the nesting site.[45]

The description of the care that Brontosaurus and her female companions exercised in choosing the site for their nests is based upon the behavior of various modern reptiles, particularly the crocodilians. As Cott has shown in his study of the Nile crocodile, the choice of a nesting site is governed by the condition of the soil, since it must be of such consistency that the crocodile can dig an adequate pit, as well as by the nearby presence of shade and by access to permanent water: "In Uganda the [nesting] sites—while differing greatly in other respects—were almost invariably on dunes overgrown with scrub, in lakeside thickets or near the fringing forest, beside isolated forest trees, or beneath an overhanging cliff. This need for shade is confirmed by the experience of Hippel (pers. comm.) and Pitman (unpub. notes)."[46] So far as the American alligator is concerned, Neill has shown that incubation of the eggs is a complicated affair; they cannot be allowed to get too hot, yet they will not incubate if they are too cool. Consequently a properly chosen site for the nest is very important for the continuation of the species. One can only suppose that the same was true for the dinosaurs.[47]

Crocodiles and alligators frequently nest in colonies, especially if they are free from outside disturbances. "Where crocodiles are entirely free from disturbance, and allowed to breed as they have doubtless done from time immemorial, they nest gregariously, the nests lying so close together that, after hatching time, the rims of the craters are almost contiguous. . . . I found many craters—the young having hatched—spaced apart, somewhat like Sandwich Terns' nests in territories, in places only three to four yards from centre to centre." [48]

The sauropod nests are pictured as being spaced somewhat more generously than those of modern crocodilians, on the assumption that such tremendously large and heavy reptiles would have required some free area around each nest in which to maneuver. Otherwise the broad feet of one female might crush the nests contiguous to her own. Moreover, the sauropod nests are indicated as having been large, in keeping with the size of the reptiles that made them. Cott shows the craters of large Nile crocodile nests as up to twelve feet in diameter. Even the smallest craters are six feet across. The American alligator constructs nests that may be five feet in diameter. Therefore it is reasonable to suppose that the sauropods made nests of considerable size; diameters of fifteen feet or more for such nests do not seem excessive.[49]

The American alligator constructs a large mound of plant debris, as much as three feet in height, perhaps in some cases even higher. An excavation is made in the top of the mound, within which the eggs are deposited and then covered. The female tends the nest during incubation, frequently wetting it with water dripping from her body after returning from a trip to a nearby pond or river. Also, she guards the nest fiercely against intruders.

The Nile crocodile, on the other hand, digs a craterlike pit in the sand. As described above, the nests may occur in communal groups. The nests are covered flush with the ground and are guarded by the females. The account of nesting by the big sauropod dinosaurs is based upon nesting in the Nile crocodile, in part because dinosaur eggs that have been discovered in various parts of the world seem to have been buried in sandy nests.

As for guarding the nests, it is here assumed that like the giant sea turtles of modern times, the sauropod dinosaurs abandoned their nests after the eggs were laid and properly covered. The crocodilians are very aggressive, carnivorous reptiles; therefore protection of the nest during incubation is an expected mode of behavior. The marine turtles are inoffensive reptiles, and it would seem that the behavior of the herbivorous sauropods may have been similar to that of the turtles.

The most abundant and best-known dinosaur eggs are those of the small horned dinosaur, *Protoceratops*, found in Mongolia, and of the sauropod dinosaur, *Hypselosaurus*, from the south of France near Aix-en-Provence. The eggs of Brontosaurus as described here are essentially *Hypselosaurus* eggs. Brontosaurus was a larger sauropod than *Hypselosaurus*; therefore the eggs might have been slightly larger. Otherwise it is reasonable to assume that they were very similar—rounded and with a pebbly surface.

The account of egg laying is based upon what we know of egg laying in crocodilians, and more particularly in the giant marine turtles, which have been observed frequently and in detail. As for the size of egg clutches, the clutches of crocodilian eggs are used as a guide. According to Cott, "Clutch counts for 775 nests [of the Nile crocodile] show an extreme range from 25 to 95 eggs, with an average of 60.4 per nest."[50] The clutch size in the Nile crocodile is in direct proportion to the size of the female.

Finally, there is some good evidence for supposing that Brontosaurus and her kind deposited the eggs in concentric circles, arranged in ascending tiers within the craterlike nest. "Where there is a sufficiency of sand, the female Nile crocodile digs a hole about two feet deep, and lays the eggs in tiers. During the process sand falls, or is shovelled, among the eggs, so that though closely-packed they are yet separated, more or less, from one another, like the currants in a cake. The upper eggs are generally about one foot beneath the surface; but in gravel they are sometimes scantily covered, and in such situations the eggs are tightly packed and the shells often dented."[51]

As additional evidence, an undisturbed nest of the ceratopsian, *Protoceratops*, was discovered in Mongolia. In it there is an inner circle of eggs, a second circle around it, and finally a part of a third circle.

Therefore the combined evidence, from modern crocodiles and one dinosaur whose nest is preserved, shows that the ruling reptiles of the Mesozoic might very well have placed their eggs within the nest in a meticulous fashion, thus affording more or less uniform warmth for each egg and some space for each hatchling to emerge.

TENTH MONTH

Confrontations

DURING MOST of their wanderings the brontosaurs went where they pleased, except when they were stopped or diverted by natural barriers such as cliffs or impassable waterways. But today, a month after and a good many miles removed from their nesting scene, they were brought to a complete halt. It was no natural obstruction that stayed their progress; it was a very large and hostile allosaur. The giant carnivore faced the brontosaurs with a demonstration so ferocious that the monstrous procession of giants stood stock-still; indeed, some of them started to back off from the raging predator with which they had come face to face.

It was a female allosaur, and her unfriendly actions were defensive rather than offensive in nature. She was guarding her nest, from which tiny babies were emerging as the eggs hatched. She had no intention of attacking the brontosaur giants in order to bring down one of them for food. Instead she was defining a boundary line beyond which no giant could advance without extreme peril to life and limb. And she was defining this line with an impressive show of force.

For many weeks she had been guarding the nest in which her eggs were cradled, for unlike the brontosaurs, she did not leave her eggs to the vagaries of weather or of nosy lizards and birds. Once having dug the nest and laid her eggs, she stayed close by to see that it was not disturbed. Nor was she alone. In the distance some other allosaurs were visible, also doing sentry duty around their nests.

Having been on guard for all of those past weeks, the allosaur mother had developed a real attachment to the nest and the area surrounding it. She had seldom left the vicinity, so she had seldom had an opportunity to hunt or to eat anything. Daily she had walked back and forth, and around the nest, occasionally wetting it with her urine, to keep it from becoming too dry and hard. Now, after a long vigil, the little allosaurs were beginning to emerge from the eggs, at the very time when the brontosaur herd managed to come upon the scene. Perhaps that was why the mother allosaur was so agitated; these intruders appeared at a very awkward moment.

Some hours before the appearance of the brontosaurs, a faint croaking was heard from within the allosaur nest. That was the signal of hatching, the call of an infant trying to break out of its prison. Within minutes, the initial voice of complaint was joined by another, and then another. The mother needed no further stimulation to initiate her response to the underground callings. With anxious haste she began to dig into the nest, using the large claws of her hands as excavating tools. Within a moment or two she exposed the first egg, already broken, with the infant partially emerged. As the egg came to light, the little allosaur made a final wriggling movement to become completely freed from the tough white shell which for so many weeks had been its home and its protection. Now it was an ever-so-small, free-ranging dinosaur, a miniature replica of the giant mother that loomed high above the nest. The little hatchling was even dwarfed by the heavy, three-toed foot of the mother allosaur. Yet in spite of its diminutive size and seeming fragility, the baby was relatively safe beneath the feet of its parent. The mother stood still so that she would not inadvertently step upon the newborn allosaur, still using the claws of her hands to dig into the nest.

She brought a second egg into view, an unbroken egg, from which issued the croaking call of a little allosaur ready to emerge. Perhaps the baby needed some help. Certainly the mother must have deemed this to be the case, because she lowered her huge head and ever so gently bit into the egg with her great front teeth. This was sufficient to start the process of hatching. The crack that she had made in the shell was enlarged by the little reptile struggling inside, and within a matter of minutes, it broke free from the encapsulating egg.

Still more croaking came from within the nest, and again the mother dug in to expose another egg. That egg, like the first one that had been brought to the surface, was already cracked, and a tiny head was beginning to thrust itself through the broken shell. Again the mother put her head down, to assist the emerging hatchling by cracking the egg still further with her front teeth.

It was at this point that she heard the sound of heavy feet on the ground and turned to see the oncoming herd of brontosaurs. Now she felt a conflict of instincts: the instinct to assist in the hatching of her eggs, and the instinct to guard her nest and her hatchlings from the intruders. The latter instinct took precedence, and she turned to face the brontosaurs, stepping high across the nest and across the babies at her feet. Of course Brontosaurus and her fellows had no intention of harming the nest or the newly hatched allosaurs; in fact, they instinctively began to retreat from the enraged mother allosaur. But, since they were ponderous beasts, their change of direction seemed all too slow to the mother allosaur, so she charged. In a confusion of long, waving necks and thrashing tails, the brontosaurs turned to flee from their implacable foe. Brontosaurus, having been at the head of the herd, as usual, was now in the rear, and as she turned her huge body, trampling the ground with her broad feet, she collided with some of the other members of the herd, also attempting to turn and run away. For two or three brief moments there was pandemonium among the brontosaurs—a scene of colliding bodies crashing together with loud thuds, the noise augmented by the bellowing of the terrified sauropods.

During this time of confusion the allosaur charged the rear of the herd, where the great bulk of Brontosaurus formed a barrier against which the allosaur had to hurl herself. She did just that, biting and clawing at the massive tail of Brontosaurus. So once again, as many months before, Brontosaurus was the object of an allosaur attack. But this time the attack was not prolonged, nor did it have very serious results. The allosaur was concerned with driving the intruding herd away; that was all. She broke off her attack almost immediately after it had been launched, and Brontosaurus escaped with a few lacerations, none very deep.

Nonetheless, the brief foray against the brontosaurs was effective. Their retreat, at first chaotic, soon became more orderly, and they began a hurried march away from the nesting site, so furiously defended by the mother allosaur. As they retreated they saw at their left two or three more female allosaurs, diligently guarding their nests and assisting with the hatching of their young, so the herd veered away from these potential trouble spots, to make a wide circuit of the area.

Brontosaurus was able within a few minutes to take her accustomed place in the van of the marching sauropods, in spite of her recently acquired wounds. They were slightly painful, but not enough to interfere with her usual activities. So she was in a position to lead the herd once again, and once more quite inadvertently, into a confrontation that was to stop the progress of the march a second time. The meeting, however, was not of serious consequence, nor did it interfere in any marked fashion with the wanderings of the herd of giants.

For the second time within the course of a morning the brontosaurs encountered dinosaurs not of their own kind; the particular reptiles in the path were stegosaurs. There was quite a gathering of the bizarre, plated, spike-tailed dinosaurs, and they were browsing across a glade that interrupted the usual continuity of the forest. Their low-placed jaws were busily engaged stripping leaves from small plants that dotted the open landscape. Their great, upstanding triangular back-plates moved through and above the foliage like multiple fins, while their spike-studded tails trailed across the ground, whispering through the plants and rattling over the bare, pebble-covered surfaces. Preoccupied with feeding, they were not at first aware of the approach of Brontosaurus and her fellows.

Then there was an awareness, and one by one, in quick succession, they turned to face the looming mass of sauropod giants. In spite of their plates and spikes, they did not seem particularly formidable to the brontosaurs. Nevertheless the herd was brought to a halt by these unexpected reptiles. Brontosaurus and her companions generally paid little attention to any stegosaurs that they met, for why should thirty-ton giants give way to dinosaurs of such modest proportions? Usually, however, encounters were with single stegosaurs, or with a very few of the plated herbivores. But today the stegosaurs were out in force, perhaps in greater numbers than Brontosaurus and the other members of the herd had ever seen. That was why the herd had been brought to a stop.

Again, but in a more peaceful and deliberate fashion than earlier in the day, the sauropod herd made a detour. It was simply a matter of not wishing to get close to those wicked, spiked tails. The brontosaurs were not enemies of the stegosaurs, but the latter were not experts in discrimination. They were apt to flail their long spikes at any intruder that ventured within range of their weapons. Thus Brontosaurus led the herd in a wide swing around the browsing stegosaurs, to enter the forest on the far side of the glade.

As the big sauropods began to browse on the trees, they were watched by several denizens of the upper branches: ancient birds, which we call *Archaeopteryx*, with thick feathers, teeth in their jaws, and long, bony tails, from which feathers radiated on each side. The birds had been disturbed by the sauropods, and they were annoyed. With harsh screams they hopped from frond to frond in a rather clumsy fashion, and finally they projected themselves into the air, to glide somewhat inexpertly to the ground. Once they had landed, they ran off into the undergrowth, still expressing their displeasure in a series of screams.

For Brontosaurus and the other members of the herd, it was the noisy finale to a morning of confrontations. The remainder of the day was peaceful enough, as were most of the days in the lives of these dinosaurian giants.

Female crocodilians guard their nests diligently and aggressively during the period of incubation. We may suppose that, in the same way, the female allosaur guarded her nest against the intrusion of the brontosaur herd. If the behavior of the carnivorous dinosaurs was anything like that of the modern crocodilians, it would seem likely that the female guarded the nest constantly, without any opportunity for feeding. According to Cott, he was informed that females of the Nile crocodile "remain on the grounds by night as well as by day. This may well be the case, for it appears almost certain that brooding females fast. . . . Stomachs of nesting females examined by me were . . . empty."[52]

There is ample evidence that female crocodilians aggressively guard their nests and their young hatchlings. Concerning the American alligator, Neill states that "under normal conditions the female will guard her nest, even against man. The guarding behavior looks like an attack, but it is not. As a person approaches the nest, the guardian female rises, turns rather slowly toward the intruder, and begins to inflate with air. If the person halts about 20 feet from the nest, the female may do nothing but hiss. If the person continues to approach, however, the reptile will lunge open-mouthed in his direction. . . . If the person turns and flees, the alligator will follow for a short distance. In other words, the guarding behavior is a series of stereotyped actions which cause the alligator's potential enemy to retreat from the vicinity of the nest."[53] As Neill further points out, "Ritualized confrontation, without actual bloodshed, is widespread among animals. . . ."[54]

Why is the female allosaur pictured as guarding her nest so fiercely, while the female brontosaurs are depicted as laying their eggs and then going away and leaving them? The parallels here are drawn between allosaurs and crocodiles, both aggressive predators, and between brontosaurs and marine turtles, both harmless nonpredators. The discoveries of sauropod eggs in southern France give the impression that they were deposited in abundance and abandoned, as are the eggs of nesting marine turtles. Why shouldn't the carnivorous dinosaurs have had nesting behavior similar to that of the most aggressive of modern reptiles—the crocodilians?

The description of the female allosaur assisting her hatchlings by digging the eggs out of the ground, and by gently cracking the shells, is based upon recent knowledge of crocodilian behavior. Again according to Cott, "The young croak as hatching time approaches, and this is the signal for the mother to uncover and so liberate her offspring. . . . Liberation by the female is necessary because, after three months' incubation and trampling, the earth above the eggs becomes compacted. At some nests opened by me the ground had to be chipped away with a knife and the eggs were set in a matrix as hard as mud-brick. Emergence through a foot or more of sun-baked and hard-packed soil would be impossible for the unaided young."[55]

Recent observations have shown that a mother crocodilian will assist the young to emerge from the egg by gently cracking it with her teeth. What is especially interesting, and most unexpected, is that the female crocodilian will take the newly hatched young in her mouth, and carry them (often several at a time) to the water, where she releases them in the shallows among abundant reeds and other water plants, which afford ample protection.[56] It is this behavioral pattern, only recently seen, and in certain respects rather atypical for a reptile, that may have given rise to some of the reports of cannibalism among the crocodilians.

As was mentioned earlier, the brain in *Stegosaurus* was remarkably small, actually and proportionately smaller than in most of the dinosaurs. It can be assumed that *Stegosaurus* was probably slow-witted, as compared with its dinosaurian cousins. Therefore, the description of the stegosaurs reacting to the sauropod herd in a predominantly instinctive fashion may be in keeping with evidence as to the development of the brain in that particular dinosaur.

The habitat and the habits of *Archaeopteryx* have been

much debated through the years. Early students of this first bird indicated that it may have been a clumsy flyer, or perhaps a gliding animal. Recently John H. Ostrom has advanced the theory that *Archaeopteryx* was not able to fly, basing his conclusions primarily on the anatomy of the shoulder girdle and the forelimb. It is his contention that the coracoid bone in the shoulder girdle is so primitive that it would not have allowed the supracoracoid muscle to act as an elevator of the wing, as it does in modern birds. Ostrom therefore says that *Archaeopteryx* was a ground-dwelling animal, using the wings somewhat in the manner of "scoops" to surround and rake in insects and other prey.[57]

But this is a weak argument. The long, flexible neck of *Archaeopteryx*, allowing wide and rapid movements of the skull, would have been sufficient for catching insects and other prey. It would appear that the large, feathered forelimbs were not particularly well suited for catching prey, based on behavior patterns in modern predators. Moreover, it seems very likely that the well-developed primary feathers on the wings of *Archaeopteryx* had an aerodynamic function. W. B. Heptonstall, in a careful evaluation of the problem, has concluded that *Archaeopteryx* was capable of gliding at an angle of 6 degrees or greater, which is sufficient for a certain amount of aerial mobility.[58]

So while Ostrom may be correct in thinking that *Archaeopteryx* could not engage in flapping flight, it seems reasonable to assume that Heptonstall is equally correct in thinking that this ancient bird could glide. In our story *Archaeopteryx* is pictured as gliding out of trees when disturbed by the sauropod herd. Perhaps *Archaeopteryx* did not fly up into the trees in the first place, but it could very well have climbed there. *Archaeopteryx* was not completely a bird but was well on its way.

ELEVENTH MONTH

The Flood

THE HERD, with Brontosaurus in the van, was cruising through the jungle in the accustomed manner, browsing from the treetops and from the forest floor, and retreating into nearby swamps during the heat of the day. The sauropods had now directed their course toward the southwest, paralleling the coast of the southern sea. Even though they were creatures of jungle and swamp, they quite frequently seemed to prefer being near the seashore; the warm, shallow waters along the strand of this tropical ocean offered a familiar retreat when they sensed the ominous presence of giant predators in their feeding grounds. They also liked to be in the vicinity of the sea because into it flowed numerous rivers, contained in broad, shallow valleys; along those rivers were abundant feeding grounds. The terrain of the river valleys near the ocean was less hilly than farther inland, an added attraction to the wandering giants.

On a bright day some time after the nest building and egg laying, the herd entered a broad valley, through the middle of which flowed a sluggish, muddy river. On each side of the river the vegetation was thick and lush—

decidedly attractive to the hungry dinosaurs. Here they paused to feed along the banks, at times walking along near the river, and at times wading through the shallows, in order to reach the plants that hung out over the water. For them it was a peaceful time, so they browsed unhurriedly. Some of their feeding was done in the river, in back reaches where aquatic plants covered the surface of the water. Here they enjoyed the advantages of both forest and marsh, because they could wade along, reaching high for the leaves of bushes and trees, and reaching low for plants growing in the water. At times some of the sauropods waded well out into the river, especially if a small island, hidden by thick vegetation, beckoned them. Their progress was marked by a scattering and fleeing of crocodiles in the water and lizards on the land.

As the day wore on, clouds began to build up to the north in towering masses. The sky became darker and there were vivid flashes of lightning. The storm on the far horizon was truly spectacular, but it was many miles away from where the dinosaurs were feeding, so they paid little attention to the blackness of the heavens and the continual rolling of thunder. They proceeded to walk and wade along the river, following it upstream in the direction of the storm. In the course of an hour or so their splashing progress brought them into a stretch of the river bounded by rather high banks. Here some of the dinosaurs were confined between the riverbanks, having failed to come out of the water downstream, where the banks were low enough for them to do so. The other members of the herd were high up on one bank of the river, paralleling their wading friends in the upstream journey. Among the dinosaurs taking the high road was Brontosaurus; she had led her division of the herd out of the water and onto dry land a half mile or so downstream, where the fringing bank was still low and gently sloping. Thus Brontosaurus and her contingent browsed through trees and bushes, while the dinosaurs in the river were wading against the current in an effort to get to a spot where they, too, could come out to join their dry-footed companions.

It was at this juncture that the flood caught the sauropods in the river. The rumbling storm on the northern horizon had grown into a cloudburst and struck the land upriver. The downpour had swelled the river until it could no longer absorb the volume of new water. The waters of the cloudburst built up into a roaring flood that came downstream, carrying everything before it. Great volumes of sand and mud were carried by the augmented waters, while on the surface uprooted trees came tumbling and turning downstream with terrifying speed. The wading sauropods were in the path of the roiling waters, and there was no way for them to escape. As the flood struck, they were knocked off their feet and were rolled over and over down the river.

All were carried downstream except two or three of the strongest sauropods. They barely managed to struggle through the pitching waves, eventually to crawl out on shore downstream, where an hour before they had disregarded the gently sloping banks. Now these banks were their salvation; they emerged from the river exhausted, and on the low ground beyond the raging water they paused for a long time to rest and recuperate.

Such welcome escape from death was not to be the lot of the remaining sauropods caught by the flood. Down the river they went, rolled and battered by the force of the water, until one by one they succumbed. Their resistance had been worn down by the inexorable force of the river so that their struggles against it grew gradually weaker and finally ceased. Now they were merely huge bodies being transported to some final resting place.

Brontosaurus and the other sauropods on the high bank were the passive and generally unemotional spectators of this drama. There was no place in the reptilian brain of the giant sauropod and her kind for compassion, for identification with the plight of the victims. She and her companions stood on the high bank watching their fellows—and that was all: As they watched, the struggling and drowning sauropods were diminished to less than giants and then to mere specks as the river carried them away, until finally they disappeared around a distant bend.

The inert forms of the victims continued to float through that day and on into the night. By then the waters began to recede, so that the giant bodies were carried at progressively slower speeds as the river rolled along. Then, one by one, the carcasses grounded on a large sandbar partially obstructing the flow of the river. And there they remained.

When the sun came up the next morning the river was flowing even more slowly than it had during the night, and the sauropod carcasses were thoroughly grounded on the sandbar. They lay in awkward attitudes, the massive bodies and the huge limbs in varied positions, with the long necks and tails stretched downstream, more or less parallel to each other, carried in that direction by the force of the flowing river.

Beneath the hot sun the great bodies began to bloat, gradually expanding from the accumulation of gases in the digestive tract. But before the bloating of the carcasses had progressed very far, they were discovered by a small group of allosaurs that appeared on the bank of the river contiguous to the sandbar. The giant predators, not at all particular about how they obtained their food, promptly waded out through the shallow water covering the bar, to attack the carrion so providentially provided for them. They were not in the least deterred by the rather second-hand nature of the meal that was spread before them in such abundance. Immediately they devoted their individual attentions to the massive bodies, now partially buried in the sand. They slashed and tore at the dead sauropods, gulping down huge gobbets of flesh as they detached pieces from the carcasses. Nor were they particular about what they severed and ate. The river water ran red about their feet.

As a result some of the bodies became partially dismem bered, although others—the larger ones especially—remained reasonably intact. Some of the smaller bones of the skeletons were destroyed—torn from their places and chewed into fragments. Others were scattered. Some bones, even though not separated from skeletons, bore the marks of the daggerlike teeth of the predators.

The feast went on until the allosaurs were sated. Then, one by one, they left the sandbar, wading to the riverbank to disappear in the jungle.

The day wore on, and the carcasses that had not been torn to pieces by the feasting predators continued to expand beneath the heat of the sun. There they remained as the sun went down and through the days that followed, becoming ever more putrid. Yet despite their deteriorating condition and the insects swarming around them, the carcasses were visited by various carrion feeders—little lizards and other small reptiles, and even some fishes that lived in the river.

The constantly flowing water, loaded with sediment, had its effect upon the inert remains of the giants. Gradually the carcasses, as they were reduced by scavengers and by the inexorable processes of decay, became buried in the sands of the bar. The tough ligaments and tendons of some preserved partial, or even complete, articulated skeletons and retained the relationships of bone to bone that had once made the powerful frames that carried the weights of gigantic bodies. So as day followed day the huge carcasses decayed and disappeared, first beneath the surface of the flowing river, finally beneath the shifting sands of the bar. That half of the herd had vanished from the land in which they had lived. Here they would remain hidden from prying eyes during millennia almost beyond counting, until in our century they would be exposed in the ancient sandbar, now a high cliff in western mountains, to be excavated by paleontologists and restored to their former appearance.

All of these happenings were lost to Brontosaurus and that part of the herd which had been on the riverbank when the flood came. They returned to their browsing. An episode had taken place in their collective lives—an episode that soon was a forgotten happening of the past.

The account of the dinosaurs caught in the flood and buried on a sandbank in the river is based largely upon an interpretation of the great dinosaur quarry at Dinosaur National Monument, on the Utah–Colorado line east of Vernal and more distantly east of Salt Lake City. Here the sediments, consisting of sandstones, and at one time horizontal, are now tilted to an angle of some 67 degrees as a result of powerful forces that caused the uplift of the western mountains. Of course all of that deformation of the earth's crust took place millions of years after the dinosaurs had been alive. Bones were discovered in these uplifted sediments in 1908. The quarry was worked for many years by parties from the Carnegie Museum of Pittsburgh, under the direction of Earl Douglass. In 1915 President Woodrow Wilson signed the bill creating Dinosaur National Monument, and within the past two decades the quarry has been developed by the National Park Service, so that now visitors can see numerous bones of dinosaur skeletons exposed in place within the rock. As a result of his unparalleled familiarity with and knowledge of the deposit, Douglass came to the conclusion that the skeletons were buried as a result of being washed down an ancient river, to be lodged upon and buried within a sandbar. There is good evidence to support his conclusions.

For example, the undersides of the skeletons are commonly composed of articulated bones; those were the bones first buried in the sand and thus held in place, even after decomposition had set in. The bones on the upper sides of the skeletons are commonly scattered, probably as a result mainly of stream action. Furthermore, the long necks and tails of the sauropod dinosaurs are strung out as if they had been so positioned by the action of the constantly flowing river.

That the rock enclosing the fossils represents an ancient sandbar is indicated by the bedding of the sandstone. Instead of being composed of nicely parallel layers, the sandstone is cross-bedded, which is to say the sandy layers are at varying angles to each other. That is the result of being deposited by shifting currents that changed in local directions, to deposit the sands in different attitudes and at different angles. Cross-beds laid down in streams are quite characteristic and can be closely analyzed. (Cross-beds also are deposited by winds, but aeolian cross-bedding is readily distinguished from aquatic cross-bedding.)

The sands that constitute the quarry at Dinosaur National Monument show varied degrees of sorting and size. Some of the cross-beds consist of very fine sands, deposited when the river was flowing sluggishly; others, of coarse sands and even of pebbles, when the rate of flow was faster, because of the action of floodwaters. Some of the scattered bones are water-worn, indicating that they had been washed down the river.[59]

At Dinosaur National Monument there are the remains of various kinds of dinosaurs of the late Jurassic Period. One is not to suppose that the skeletons were deposited as the result of a single flood. They very probably represent an accumulation of bodies that extended through many years. Our story of the sauropods being caught in a flood and washed down to be buried on a sandbar might be one incident in the history of Dinosaur National Monument.

Whether or not bones from the skeletons at the Monument were scattered by predators, or chewed upon, we know that such things did happen. They happen often enough today—on the African veldt, for example. At Bone Cabin Quarry in Wyoming, another great deposit of dinosaur skeletons excavated before the turn of the century by the American Museum of Natural History in New York, a skeleton of a sauropod was discovered of which numerous tail vertebrae show toothmarks of the right size and spacing to have been made by a large *Allosaurus*, chewing on a carcass. Perhaps the sauropod was a victim of predation; perhaps the allosaur was feeding upon carrion. Whatever the case, it is obvious that a large carnivorous dinosaur had bitten off the spines of the vertebrae in the sauropod tail.

To throw additional light on the matter of dinosaur burials, mention may be made of the truly gigantic quarrying operations carried on by the Berlin Museum during the early years of this century at Tendaguru, Tanzania. Here were found numerous skeletons of late Jurassic sauropods and other late Jurassic dinosaurs—the equivalents of the dinosaurs found in North America.

It has been postulated from the evidence of the sediments that the dinosaurs were buried at Tendaguru in a lagoon near the sea, at the mouth of a river. The reason for reconstructing such a locale is that at Tendaguru there is an alternation of fresh water and marine deposits. That is the result of the river's bringing in its sediment from the land and of the sea's washing in over a bar from the opposite direction. The dinosaurs were washed down by the river to be buried near its mouth, at the edge of the sea.

According to John Parkinson, "Doubtless many animals were caught and entombed in the lagoons themselves, many more swept seawards when the sandy barriers shutting out the sea were broken through, and mud, silt, and bones plastered out, fanwise, on to the shell-banks of the shore."[60]

Evidence concerning the burial of big dinosaurs is often at hand and clearly written in the rocks. Our story, then, is not fanciful. It is a re-creation of some things that happened many millions of years ago and were recorded within the sandy sediments of a very ancient river. It is a story of tragedy and death, a story of the kind that has taken place over and over again since time immemorial. From such misfortunes to living things in ages past, the paleontologist today is able to obtain glimpses not only of what life was like in those distant days, but also of happenings that affected those beings of long ago. The record in the rocks often conveys a sense of immediacy that lends life and drama to the long-inert bones.

TWELFTH MONTH

The Hatchlings

IT WAS NOW more than a month since the tragedy at the river, and Brontosaurus, with the diminished herd, was back in territory that seemed familiar. As the dinosaurs continued on their way, the landscape became increasingly recognizable, as well it might, for the herd was coming back to the region where, more than three months ago, some of their members had industriously dug their nests and laid their eggs. The sauropods had traveled in a circle, as was so often their custom—a circle that covered many miles during many days, even though it was not so extended as the long ellipse they had traversed in their journey between the southern and northern seas. Going out and coming back during their feeding excursions was for them a pattern of living; the pull of well-known surroundings was a strong and determining force in their lives. Once again they were returning home.

Although the scene presented a friendly vista in which there were landmarks recognizable from earlier associations, there suddenly appeared some new dinosaurs which they had never seen before. The dinosaurs were not strange—they were of their own kind—but they were unfamiliar. They were juvenile sauropods, hatched several years previously, living by themselves during the intervening years, and now come for the first time to join and augment the herd. Did they belong to this herd? Had they been born from eggs laid by Brontosaurus and her female companions? Perhaps; perhaps not. Whatever their origins, they were now asking to join this herd, because there was, at that moment, a chance meeting.

There were several of them, perhaps a half dozen or so, and they approached their superiors with proper caution. Well might they be cautious, because the members of the returning herd, even the smallest of them, loomed impressively before the eyes of these newcomers. The juveniles that had been traveling with the herd for so many months were advanced youngsters, perhaps half as long from head to tail as their elders, but these potential recruits were much smaller. Even the largest among their little coterie was scarcely a fourth of the length of the adult sauropods.

Yet even so, these relatively small sauropods were reptiles of no mean proportions, with lengths of twelve to fifteen feet or so, with weights of several tons apiece. Where had they come from? Where had they been? In immediate terms, they had come out of the forest ahead of the advancing herd. Where they had been before that is a mystery. But since fossils of very small sauropods have not been found with the other fossils, the young must have been far removed from Brontosaurus and her herd.

Here they were, approaching the giant sauropods in a bid for acceptance. As they drew close they stopped, as did the members of the herd, and both groups eyed each other warily. Then, as Brontosaurus and her companions remained standing, the small sauropods resumed their approach, slowly and with hesitating steps. Finally they came within touching distance, to make their acquaintance with the members of the herd. Heads were moved up and down and from side to side on long, swaying necks, and there were close inspections on the part of all concerned. The little ones looked closely at the giants, and the giants in turn critically examined these newly arrived sauropods. The ceremony went on for some time, with much moving about and mingling of the curious dinosaurs. Finally the newcomers were accepted by the members of the herd; they joined the larger group and the enlarged herd moved on its way.

In this way the herd became more varied; it now consisted of adults of various ages, of well-advanced juveniles, and of those smaller juveniles. As they traveled the newcomers joined their somewhat larger juvenile friends in the center of the group. As usual, the large sauropods traveled on the fringes of the advancing herd, forming a wall of massive bodies to protect the smaller dinosaurs. As usual, too, a few of the largest and most experienced sauropods, with Brontosaurus in the lead, constituted the foremost elements in this marching mass of giants. It was a well-organized brigade of walking, browsing reptiles.

It was necessary for the new members of the herd to keep up with the rest of the dinosaurs, to maintain their place among their newfound friends. That they did, and for the most part without difficulty. The sauropods were

As the hatchlings emerged into the bright light of the morning they immediately set off in the direction of the fringing forest, each baby sauropod running as fast as its short legs would carry it toward havens of thick ferns. So fast was the escape of the newly hatched brontosaurs from the sites of their nests to the protective undergrowth, it would seem as though each tiny dinosaur disappeared almost as rapidly as it had appeared out of the ground. The instinct among the baby dinosaurs for running toward heavy shade was so overwhelming that they formed a broad front of scampering reptiles, all running toward the forest. It was so strong that the herd of sauropod giants intervening between many of them and the forest was in no way a deterrent to their frantic progress. The scampering babies ran past and among the broad feet of the big sauropods, as if these were the bases of so many trees standing along the routes of their first short journeys.

There was a reason for the speed with which the hatchlings ran toward the sheltering forest. Between their nests and the protective undergrowth they were extremely vulnerable to onslaughts from various hungry predators, which suddenly appeared in abundance and variety. For many small, carnivorous dinosaurs the emerging sauropods furnished abundant feasting. These agile predators were joined by others: large lizards from the forest, small crocodiles from the nearby sea, and large flying reptiles, pterosaurs, that swooped down from the air to pick up such ready and delectable morsels. Many of the hatchlings never made it to the protection of the undergrowth, yet many of them did.

Many of the survivors that reached the forest subsequently would fall prey to forest predators, such as the smaller carnivorous dinosaurs and the lizards, but in the natural course of things some of them would survive. These survivors were destined to lead a most secretive life for many years, eating the soft vegetation of the forest floor, gathering energy from the green plants, and growing ever larger as the months and the years passed by.

Eventually some of the little dinosaurs would attain such size that they could move freely through the jungle with only the large predatory dinosaurs, such as *Allosaurus*, to be feared. Those juveniles would seek the protection of some wandering sauropod herd, just as the little group of adolescent sauropods had joined the herd led by Brontosaurus. So the cycle would have been completed and the population of sauropods would have been replenished, to fill the vacancies that had taken place as a result of predation by carnivorous giants, and of accidents, such as falls or floods.

But Brontosaurus did not appreciate the replenishment of her kind by the emergence of the hatchlings. She had dug the nest and laid the eggs, and with that her task was done. Now it was time for the herd to move on, time for the giants to continue with their browsing. They turned aside from the edge of the nesting ground, to reenter the forest in search of sustenance. With slow and majestic tread they went on their way. Life in the world of sauropods had begun anew before their eyes, but for them the immediate concern was to go on living as they had during long years past. Brontosaurus reached up into a tall tree-fern to strip it of some of its foliage. All around her the other sauropods were browsing, high in the trees and low in the undergrowth. Daily routine for the giants had not changed; it was continuing its course as it had for untold years in the past and as it would for untold years in the future.

The incident of the sauropod herd being joined by several juveniles coming out of the forest is fanciful, but perhaps in line with what might have happened. It is very probable that after hatching, young sauropod dinosaurs spent several years in some sort of isolation, until they were large enough to join the adults. The sauropod footprints found in Texas reveal a large company of these dinosaurs, consisting of adults and subadults. There are no indications of truly small sauropods in their midst. Moreover, fossils of little sauropods, newly hatched, are nonexistent. A skeleton of a juvenile *Camarasaurus*, about sixteen feet in length, in the Carnegie Museum in Pittsburgh, is one of the smallest sauropods known. Truly small dinosaurs of this type do not occur in the fossil record. Nevertheless, it is supposed that very young sauropods were small, because they had to emerge from eggs that necessarily were limited in size. The dinosaur eggs from southern France, presumably of the sauropod *Hypselosaurus*, are about twice the size of ostrich eggs.

Therefore, there is reason to think that perhaps the very young sauropods lived in an environment separate from that of the elders, which is the case among modern crocodilians. If this were true, why should not juvenile sauropods join the adults when they were large enough to travel with the herd? It seems likely.

The hatching of the baby sauropods and their flight toward the protection of the forest is based upon our knowledge of the egg hatching of modern crocodilians, and especially of marine turtles.

Among the marine turtles the mother abandons the nest after the eggs are laid and the nest covered, so that the babies must make their way into the world unassisted. They emerge from the shells, dig out of the nests, and then make a concerted dash across the beach for the sea, running a gauntlet of predatory seabirds, mammals, and frequently large crabs. It is believed that the hatchling turtles go toward the ocean for refuge because the light of the sky is brighter there than it is over the land: they are led on by positive phototropism.

In this story it is imagined that the little sauropods might have been directed toward the forest by an opposite instinct: a desire to run toward shade. Why not? It seems likely that the little reptiles would have sought the shade of the forest for protection in the undergrowth, just as modern baby sea turtles seek the bright light over the sea for protection in the water. And as is the case with the baby marine turtles, the little sauropods probably had to run from many predators. They would have been fair game for the small carnivorous dinosaurs of their time, such as *Ornitholestes*, and for big lizards. At the time of this story, birds were in the initial stages of their evolution. Perhaps they were sufficiently advanced to have fed upon young sauropods; however, for the sake of a conservative viewpoint, they are not brought into the drama. Instead, it is supposed that the struggling little sauropods were subjected to aerial attack from pterosaurs, or flying reptiles, which were abundant and varied at that time.

The supposition that baby sauropods went into hiding and remained apart from the adults until half grown is based upon our knowledge of modern crocodilians, as studied by Cott: "A few days after hatching the young lead a life of seclusion, shunning both the basking grounds and the open water, and seeking sanctuary from numerous enemies, including members of their own species, among the papyrus, in sudd, weedy shallows, backwaters, or in isolated pools—occasionally at some distance inland. A striking instance of this segregation of the young from their elders was seen at Magungu, where yearlings occupied a chain of shallow, weed-choked meres a mile back from the river. . . . By their second year they have vanished, and their disappearance illustrates a curious and little understood phenomenon. For with growth, there follows a period when juveniles mysteriously disappear from the scene in all habitats. One sees those that have recently hatched, and again crocodiles of five feet and upwards. But although the intervening sizes can be shot in the shallows at night, their diurnal whereabouts remains an enigma. . . . In areas, where under protection, crocodiles are still abundant . . . the apparent absence of small crocodiles seems all the more remarkable. All that one can certainly say is that between the ages of about two and five years crocodiles go into retreat; and since they can have few enemies other than larger individuals of their own kind, it is probable that this cryptic behaviour has been forced upon them by the habit of cannibalism." [61]

One can hardly suppose that such vegetarians as the sauropods were cannibalistic. Nonetheless, since there is no evidence for very young sauropods in the fossil record, it is logical to suppose that perhaps they, too, like young crocodiles, went into hiding, especially for protection from the various predators that existed at the time.

Finally, the sauropods' unconcern for the hatchlings is based upon the supposition that, like modern marine turtles, they did not guard the nest and consequently had no instinctive feelings for the young reptiles that came out of the nest. Brontosaurus and her kind survived by reason of the multiplicity of young produced, a few of which each year reached adulthood in spite of heavy mortality along the way.

NOTES
SELECTED AND ANNOTATED BIBLIOGRAPHY
INDEX

Notes

1. In our modern world the *ectothermic* vertebrates are the fishes, amphibians, and reptiles, whose body temperatures are essentially dependent upon the temperatures of their environments. The *endothermic* vertebrates are the birds and mammals, whose physiological mechanisms for maintaining body temperatures are generally independent of environmental temperatures.
2. The dates assigned to geologic events in millions of years are based upon sophisticated studies on the rates of decay of certain radioactive elements. Thus the ratio in some rocks of uranium to lead (lead is the end-product of the breakdown of uranium, for which the rate of decay has been determined), or of strontium to rubidium, is a guide—or at least a clue—to the age of such rocks. These determinations contain a relatively small margin of error, and they are being revised from time to time, as new data become available. As of now some widely accepted figures for the Mesozoic Era are as follows:

	Ended (in millions of years ago)	Began (in millions of years ago)	Duration (in millions of years)
Cretaceous Period	65 ± 2	135 ± 5	70
Jurassic Period	135 ± 5	190 ± 5	55
Triassic Period	190 ± 5	225 ± 5	35

3. The name Brontosaurus (without italics) is here being used to designate the heroine of this story. *Brontosaurus* (with italics) has long been employed as the name for one genus of sauropod dinosaurs; the sauropods—great, heavy-bodied, long-necked, long-tailed reptiles—are the giants among the dinosaurs. But it turns out that, according to the International Rules of Zoological Nomenclature, the proper name for this dinosaur is *Apatosaurus*. So be it. For our story we will stick to Brontosaurus as the name for one particular individual of the genus *Apatosaurus*.
4. The figure of thirty tons for a large *Apatosaurus* is not a mere guess. Nor is the figure of three tons for *Allosaurus*, cited on page 6. Some years ago I attempted to determine the live weights of various dinosaurs by measuring the volumetric displacements of a series of carefully executed, carefully scaled models. The results obtained were multiplied by the appropriate factor in each case. From studies of certain modern reptiles it was assumed that the specific gravity in the dinosaurs was probably about 0.9. (Most vertebrates float in water; the specific gravity of an alligator, for example, is 0.89.) Multiplying the volumes obtained by 0.9, for the specific gravity, the probable weights of the dinosaurs were determined. See Edwin H. Colbert, "The Weights of Dinosaurs," *American Museum Novitates*, number 2076 (1962), pages 1–16.
5. *Allosaurus* is the proper generic name, although some students claim it should be *Antrodemus*. However that may be, Allosaurus (without italics) will be the name for this individual carnivorous dinosaur, as well as for other dinosaurs of that kind that appear later in the story.
6. The absence of an inner claw in the forefoot prints made by the Texas sauropod indicates that it was perhaps a genus different from *Apatosaurus*. The nature of the Texas tracks has been discussed in a recent paper by Wann Langston ("Nonmammalian Comachean Tetrapods," *Geoscience and Man*, volume 8, 1974, pages 77–102). He cites the Lower Cretaceous sauropod, *Pleurocoelus*, stating that "it is tempting to associate the Glen Rose sauropod tracks with this taxon *Pleurocoelus*" (page 96). The de-

scribed species of *Pleurocoelus* from eastern North America, specifically from Maryland and the District of Columbia, are much smaller than the dinosaurs that made the Glen Rose and Bandera tracks. As Langston says, "their poor ossification and open neurocentral sutures suggest immaturity" (page 86). Bones have been found in Texas which Langston believes may represent *Pleurocoelus*, and they are considerably larger than the bones described from Maryland. Perhaps, therefore, the association of these bones with the Texas trackways is justified. Whatever dinosaur made the huge prints at Glen Rose, it was sufficiently close to *Apatosaurus* in size and structure to serve as the basis for our story. The same is true for the Glen Rose carnosaur tracks, vis-à-vis *Allosaurus*.

7. R. T. Bird, "A Dinosaur Walks into the Museum," *Natural History*, volume 47 (1941), pages 75–81.
8. It has recently been argued by Robert T. Bakker that the giant sauropods could even extend their high probing among the treetops by standing on their hind legs ("Dinosaur," *McGraw-Hill Yearbook of Science and Technology*, New York: McGraw-Hill, 1972, pages 179–81).
9. The figure for the amount of food consumed daily by an adult elephant is a general estimate based on various sources. At the Philadelphia Zoological Garden adult elephants (of both species) are fed two bales of hay daily; each bale weighs between fifty and sixty pounds. That is, of course, dry food and is supplemented by a large intake of water. In addition, each elephant is fed a bucket of beet pulp, one-quarter bucket of alfalfa pellets, and a quart of crushed oats. These figures were supplied through the kindness and good offices of Dr. Roger Conant, former director of the Philadelphia Zoological Garden.

   Francis G. Benedict, in his study of the physiology of the elephant, showed that a female Asiatic elephant weighing 3,000 kilograms (or about 6,600 pounds) consumed daily 50 kilograms or more than 110 pounds of hay daily (again dry food), plus several kilograms of bread, oats, and bran. That elephant drank an average of 140 kilograms (308 pounds) of water daily (*The Physiology of the Elephant*, Carnegie Institute, Washington, publication number 474, 1936).

   R. M. Laws, studying wild elephants in Africa, concluded that they consumed about 6 percent of their body weight daily. A large bull weighing 5,000 kilograms (11,000 pounds, or 5½ tons) would thus eat about 300 kilograms (about 660 pounds) of food daily. That was natural grass, browse, and herbs, containing a considerable amount of water. Some 300 kilograms of such food would be the equivalent of about 75 kilograms (about 165 pounds) dry weight ("Elephants as Agents of Habitat and Landscape Change in East Africa," *Oikos*, volume 21, 1970, pages 1–15).

The figures for the tortoise were provided by J. Kevin Bowler, curator of reptiles at the Philadelphia Zoological Garden, through the kindness of Dr. Roger Conant, former director of that institution.

10. Paul Keen, *South African Journal of Science*, volume 71 (1975), pages 134–35.
11. R. T. Bird, "Did *Brontosaurus* Ever Walk on Land?," *Natural History*, volume 53 (1944), pages 61–67.
12. Ibid., page 65.
13. John H. Ostrom, "Were Some Dinosaurs Gregarious?," *Palaeogeography, Palaeoclimatology, and Palaeoecology*, volume 11 (1971), pages 287–301.
14. C. C. Albritton, "Dinosaur Tracks near Comanche, Texas," *Field Laboratory*, volume 10 (1942), pages 161–81.
15. James V. Warren, "The Physiology of the Giraffe," *Scientific American*, volume 231 (1974), pages 96–105.
16. The idea that the flying reptiles might have had an insulating coat of some sort goes back many years—well into the last century. It was for long, however, an idea based largely upon conjecture. In 1927 Ferdinand Broili of Munich published a paper in which he claimed to have seen indications of fur or hair on a Jurassic pterosaur found in Germany ("Ein Rhamphorhynchus mit Spuren von Haarbeckung," *Sitzungberichte der Akademie der Wissenschaften 30 München*, 1927, pages 49–67, plates 4–7).

    But many authorities considered that what Broili saw were artifacts in the specimen. Then in 1971 A. G. Sharov published on a pterosaur found in Russia, in which the evidence of a hairy covering is unmistakable and beyond dispute ("New Flying Reptiles from the Mesozoic Deposits of Kazakhstan and Kirgizia," *Current Problems of Paleontology*, Academy of Science, U.S.S.R., Transactions of the Paleontological Institute, volume 130, 1971, pages 104–13 [in Russian]). He named the fossil *Sordes pilosus*, "hairy devil."
17. The famous ancestral bird *Archaeopteryx lithographica*, known from five specimens, comes from the Upper Jurassic Solnhofen limestone of southern Germany. Birds of that type may very well

have been present in North America during the late Jurassic Period, but not preserved in the fossil record because of the nature of the Morrison sediments. The Solnhofen limestone is an extremely fine-grained limestone, formerly much used in lithography. Fossils preserved in this rock show the finest details—thus imprints of the feathers are preserved in specimens of *Archaeopteryx*. (Were it not for such imprints, that particular form might have been classified as a reptile.) The Morrison beds consist of sandstones and siltstones of much coarser nature. So there is good reason to think that this ancestral bird might have been a member of the Morrison fauna.

18. Iain Douglas-Hamilton and Oria Douglas-Hamilton, *Among the Elephants* (New York: Viking, 1975), page 214.
19. Wilfred T. Neill, *The Last of the Ruling Reptiles: Alligators, Crocodiles, and Their Kin* (New York and London: Columbia University Press, 1971), page 225.
20. Edwin H. Colbert, Raymond B. Cowles, and Charles M. Bogert, "Temperature Tolerances in the American Alligator and Their Bearing on the Habits, Evolution, and Extinction of the Dinosaurs," *Bulletin of the American Museum of Natural History*, volume 86 (1946), article 7, pages 327–74.
21. Ibid.
22. John H. Ostrom, "The Ancestry of Birds," *Nature*, volume 242, number 5393 (1973), page 136.
23. John H. Ostrom, "Terrestrial Vertebrates as Indicators of Mesozoic Climates," *Proceedings of the North American Paleontological Convention*, September 1969, Part D, pages 347–76.
    Alan Feduccia, "Dinosaurs as Reptiles," *Evolution*, volume 27 (1973), pages 166–69.
    John H. Ostrom, "Reply to 'Dinosaurs as Reptiles,' " *Evolution*, volume 28 (1974), pages 491–93.
    Robert T. Bakker, "Dinosaur Physiology and the Origin of Mammals," *Evolution*, volume 25 (1971), pages 636–58.
    The case for endothermism in dinosaurs has been polemically supported in Adrian J. Desmond, *The Hot-Blooded Dinosaurs* (New York: Dial Press / James Wade, 1976). As one reviewer has noted, "Desmond's thesis is a victim not so much of the information explosion as of a dogmatic adherence to an extremist view" (Richard J. Wassersug, "Dinosaur Biology," *Science*, volume 193, 1976, page 44).
24. Albert F. Bennett and Bonnie Dalzell, "Dinosaur Physiology: A Critique," *Evolution*, volume 27 (1973), page 170. The article

is a careful analysis and criticism of Bakker's paper of 1971.

25. Robert T. Bakker, "The Superiority of Dinosaurs," *Discovery*, volume 3 (1968), pages 11–22.
26. R. McN. Alexander, "Estimates of Speeds of Dinosaurs," *Nature*, volume 261, number 5556 (1976), pages 129–30.
27. Robert T. Bakker, "Dinosaur Renaissance," *Scientific American*, volume 232, number 4 (1975), pages 58–78.
28. Armand de Ricqlès, "Recherches paléohistologiques sur les os longs des tétrapodes. 1. Origine du tissu osseux plexiforme des dinosauriens sauropodes," *Annales de Paléontologie (Vertébrés)*, volume 54 (1968), pages 133–45, plate.
29. Donald H. Enlow and Sidney O. Brown, "A Comparative Histological Study of Fossil and Recent Bone Tissues. Part II," *Texas Journal of Science*, volume 9 (1957), pages 186–214.
30. Armand de Ricqlès, "L'histologie osseuse envisagée comme indicateur de la physiologie thermique chez les Tétrapodes fossiles," *Compte Rendu Académie Sciences Paris*, volume 268 (1969), page 785.
31. Armand de Ricqlès, "Evolution of Endothermy: Histological Evidence," *Evolutionary Theory*, volume 1 (1974), pages 51–80.
32. See James Gray, *Animal Locomotion* (London: Weidenfeld and Nicolson, 1968).
    Bobb Schaeffer, "The Morphological and Functional Evolution of the Tarsus in Amphibians and Reptiles," *Bulletin of the American Museum of Natural History*, volume 78 (1941), article 6, pages 345–472.
33. During the first two decades of this century there was a considerable debate as to the pose in sauropod dinosaurs. Some paleontologists thought that the sauropods affected a sprawling attitude, like some modern lizards, or crocodiles (when resting). Others contended that the sauropods walked with the limbs rather straight and with the feet well beneath the body. Some authorities thought that the sauropods could not come out on land because of their great weight; they were of necessity confined to the water, for support of their massive bodies. The Texas trackways show quite conclusively that the giant sauropods walked, like elephants, with the right and left feet close together, beneath the body.
34. Bakker, "Dinosaur," pages 179–81.
35. Douglas-Hamilton and Douglas-Hamilton, *Elephants*, pages 73–74.
36. Angus Bellairs, *The Life of Reptiles*, volume 2 (London: Weidenfeld and Nicolson, 1969), page 343.

37. Ibid.
38. Ibid., pages 343–44.
39. Neill, *Ruling Reptiles*, page 255.
40. Figures for brain and body weights in various modern reptiles as well as in certain birds and mammals are given in Bellairs, *Life of Reptiles*, page 333.
41. Roy L. Moodie, *Palaeopathology* (Urbana: University of Illinois Press, 1923). Moodie illustrates the occurrence of *spondylitis deformans* in several tail vertebrae of *Diplodocus*.
42. Neill, *Ruling Reptiles*, page 206.
43. Hugh B. Cott, "Scientific Results of an Enquiry into the Ecology and Economic Status of the Nile Crocodile (*Crocodilus niloticus*) in Uganda and Northern Rhodesia," *Transactions of the Zoological Society of London*, volume 29, part 4 (1961), page 267.
44. Neill, *Ruling Reptiles*, page 207.
45. See Bellairs, *Life of Reptiles*, page 419, for remarks concerning sperm storage in reptiles; also Neill, *Ruling Reptiles*, pages 208–19, for courtship and nesting of the alligator.
46. Cott, "Nile Crocodile," page 270.
47. Neill, *Ruling Reptiles*, pages 212–17.
48. Cott, "Nile Crocodile," page 271.
49. Ibid., page 272, a map showing the distribution of seventeen craterlike crocodilian nests contained within an area of about five hundred square yards, or about one-tenth of an acre. One nest is about twelve feet in diameter and four others are each about ten feet in diameter. The smallest nests have diameters of about six feet. Twelve of the nests have their rims either touching, or within a foot or so of one another.
50. Ibid., page 275.
51. Ibid., page 273.
52. Ibid., page 275.
53. Neill, *Ruling Reptiles*, pages 213–14.
54. Ibid., page 215.
55. Cott, "Nile Crocodile," page 276.
56. Anthony C. Pooley and Carl Gans, "The Nile Crocodile," *Scientific American*, volume 234, number 4 (1976), pages 114–24.
57. John H. Ostrom, "Some Hypothetical Stages in the Evolution of Avian Flight," *Smithsonian Contributions to Paleobiology*, number 27 (1976), pages 1–21.
58. W. B. Heptonstall, "Quantitative Assessment of the Flight of *Archaeopteryx*," *Nature*, volume 228 (1970), pages 185–86.
59. Theodore E. White, *Dinosaurs at Home* (New York: Vantage,

1967). An account of the manner in which the numerous skeletons were accumulated and deposited at Dinosaur National Monument is presented on pages 177–80.

60. John Parkinson, *The Dinosaur in East Africa* (London, A. F. and G. Witherby, 1930), pages 127–29.
61. Cott, "Nile Crocodile," pages 277–78.

Selected and Annotated Bibliography

MANY PUBLICATIONS WERE CONSULTED during the preparation of this book; no attempt will be made to list them all here. Some of them are cited in the Notes, and of those a few are included in the following selective list. As is readily apparent, the bibliography that follows is restricted to certain particularly pertinent works, some of them containing extensive bibliographies of their own. They contain the core of information bearing in one way or another upon *The Year of the Dinosaur*. But many items upon which the story has been based have been gleaned from sources too numerous for practical inclusion here.

Dinosaurs

COLBERT, EDWIN H. *Dinosaurs: Their Discovery and Their World.* New York: E. P. Dutton, 1961. xiv plus 300 pages.

A general book dealing with the various aspects of dinosaurian discoveries, evolution, adaptations, distribution, classification, and extinction.

———. *The Age of Reptiles.* London: Weidenfeld and Nicolson, 1965; New York: Norton, 1965. xiv plus 228 pages.

A book concerned with the late Paleozoic and Mesozoic eras, when reptiles were dominant. Much attention is given to the dinosaurs.

EDMUND, GORDON. "A Search for Dinosaur Fodder." *Rotunda; The Bulletin of the Royal Ontario Museum*, volume 6, number 1 (1973), pages 39–44.

The article here cited contains excellent color photographs of modern tropical and subtropical plants of the types that may have been eaten by herbivorous dinosaurs.

HOTTON, NICHOLAS, III. *Dinosaurs*. New York: Pyramid, 1963. 192 pages.

A very useful paperback volume, dealing with the evolution and the adaptations of the dinosaurs.

ŠPINAR, ZDENĚK V. *Life before Man*. Illustrated by Zdeněk Burian and A. Beněsová. Translated by Margot Schierlová. London: Thames and Hudson, 1972. 228 pages.

This large picture book is especially valuable because of the authoritative and carefully executed restorations of dinosaurs and the environments in which they lived, made by Burian under the guidance of the late Professor Joseph Augusta, Z. V. Špinar, and V. Mazák.

SWINTON, W. E. *The Dinosaurs*. New York: John Wiley, 1970. 331 pages.

A revised edition of a classic work, first published in 1934. It is especially valuable for its treatment of dinosaurs found in Europe. There is a rather extensive bibliography.

White, T. E. *Dinosaurs at Home*. New York: Vantage, 1967. 232 pages.

This book deals with dinosaurian evolution, distribution, and extinction, with particular emphasis on the dinosaurs found at the Dinosaur National Monument, Utah.

Sauropod and Carnosaur Dinosaurs

COOMBS, WALTER P., JR. "Sauropod Habits and Habitats." *Palaeogeography, Palaeoclimatology, Palaeoecology*, volume 17 (1975), pages 1–33.

The author attempts to analyze the natural history of the sauropods, as interpreted from the evidence of anatomy, the use of modern analogs, taxonomic diversity, footprints, and sedimentology. Some twenty-five conclusions are reached concerning the adaptations and habits of the sauropods. A considerable bibliography pertaining to sauropods is included.

GILMORE, C. W. "Osteology of the Carnivorous Dinosauria in the

United States National Museum, with Special Reference to the Genera *Antrodemus* (*Allosaurus*) and *Ceratosaurus*." *Bulletin of the United States National Museum*, Bulletin 110 (1920), pages 1–154, 36 plates.

———. "A Nearly Complete Articulated Skeleton of *Camarasaurus*, a Saurischian Dinosaur from the Dinosaur National Monument, Utah." *Memoirs of the Carnegie Museum*, volume 10 (1925), pages 347–84.

———. "Osteology of *Apatosaurus*, with Special Reference to Specimens in the Carnegie Museum." *Memoirs of the Carnegie Museum*, volume 11 (1936), pages 175–300.

These monographs on sauropod dinosaurs present detailed descriptions of the bony anatomy and adaptations of those reptiles. The monograph on the carnivorous dinosaurs is included here because of its information concerning *Allosaurus*.

Hatcher, J. B. "*Diplodocus* (Marsh): Its Osteology, Taxonomy and Probable Habits with a Restoration of the Skeleton." *Memoirs of the Carnegie Museum*, volume 1 (1901), pages 1–61.

Another memoir on a sauropod dinosaur, useful in conjunction with the monographs by Gilmore.

Osborn, H. F., and Mook, C. C. "*Camarasaurus*, *Amphicoelias*, and Other Sauropods of Cope." *Memoir of the American Museum of Natural History*, new ser., volume 3 (1921), pages 249–387, plates 60–95.

This impressive work includes detailed descriptions and elaborate plates showing the details of sauropod anatomy. There are some excellent restorations.

Ostrom, John H., and McIntosh, J. E. *Marsh's Dinosaurs: The Collections from Como Bluff.* New Haven: Yale University Press, 1966. xiv plus 64 pages, 65 plates.

Ostrom and McIntosh have published many of the plates, heretofore unpublished, originally prepared for O. C. Marsh, the founder of Yale Peabody Museum and an early student of dinosaurs, and have written a most useful text to go with them, describing the history of field work at Como Bluff, Wyoming. Excellent illustrations of sauropod bones are included.

Crocodilians and Other Modern Reptiles

Bellairs, Angus. *The Life of Reptiles.* London: Weidenfeld and Nicolson, 1962. Volume 1, pages i–xi, 1–282; volume 2, pages 283–572.

For anyone attempting to reconstruct the life of long-extinct reptiles, these volumes are invaluable. The characters, distribution, anatomy, physiology, habits, reproduction, ecology, and classification of modern reptiles are described and discussed. There is an extensive bibliography.

COTT, HUGH B. "Scientific Results of an Inquiry into the Ecology and Economic Status of the Nile Crocodile (*Crocodilus niloticus*) in Uganda and Northern Rhodesia." *Transactions of the Zoological Society of London*, volume 29 (1961), pages 215–337.

MINTON, SHERMAN A., AND MINTON, MADGE RUTHERFORD. *Giant Reptiles*. New York: Charles Scribner's Sons, 1973. xiii plus 345 pages.

NEILL, WILFRED T. *The Last of the Ruling Reptiles: Alligators, Crocodiles, and Their Kin*. New York: Columbia University Press, 1971. xvii plus 486 pages.

All three of these works are essential in helping one to understand the dinosaurs, specifically because crocodilians are the living reptiles most closely related to the rulers of Mesozoic time. A large portion of the book by the Mintons is devoted to the crocodilians. Cott's work is based upon much observation of a single species in the field. Neill's book is a compendium from many sources. The three publications throw much light on crocodilian habits and behavior. All have good bibliographies; the one in Neill is very extensive.

Elephants

BENEDICT, FRANCIS G. *The Physiology of the Elephant*. Carnegie Institution of Washington, publication number 474, 1936. vii plus 474 pages.

DOUGLAS-HAMILTON, IAIN, AND DOUGLAS-HAMILTON, ORIA. *Among the Elephants*. New York: Viking, 1975. 285 pages.

These two interesting works are included because an understanding of elephants, particularly with regard to problems of giant size, is helpful to an understanding of sauropod dinosaurs. Benedict's monograph is a careful physiological study of an elephant in a specially constructed laboratory. The book by the Douglas-Hamiltons recounts the experiences and results of living among wild elephants in Africa and observing them at very close range.

Index

Albritton, Claude C., 156
  analysis of sauropod tracks by, 27–28
Alexander, R. McN., 158
alligators
  delay between mating and nesting of, 114
  nest of, 115–16
  temperature tolerances of, 58
allosaur female
  aiding of hatchlings by, 120–22
  charging of sauropod herd by, 122–23
  guarding of nest by, 119–20
allosaurs
  attack on sauropod herd by, 89–92
  comparison between crocodiles and, 126
Allosaurus, xiii, 147
  distribution of, 82
  as proper generic name, 154
  toothmarks of on sauropod bones, 138
Allosaurus, 76
  attack on Brontosaurus by, 7–8
  description of, 5–7
  footprints of, 11
American Museum of Natural History, New York, 9, 138
Antrodemus, 154
Apatosaurus, xii, xiii
  determination of weight of, 154
  distribution of, 82
  size of, 25
  teeth of, 22
Archaeopteryx
  ability of to fly, 128–29
  behavior of, 127–29
  description of, 40–41, 125
  fossils of, 156–57
arthritis in dinosaurs, 103

Bakker, Robert T., 158
  on endothermism in dinosaurs, 60–61
  on sauropod posture, 155

Bandera, Texas, 26, 72, 80
Barosaurus, 35
basking by reptiles, 47
behavior and bones, vii–ix
behavior in dinosaurs, ix–x
Bellairs, Angus, 158, 159, 163
  on vision in reptiles, 92–93
Benedict, Francis G., 155, 164
Bennett, Albert F., 157
  and Bonnie Dalzell, on relation of posture to body temperatures, 61
Bird, R. T., 155, 156
  discovery and excavation of sauropod tracks by, 9–11
  discovery of tracks at Bandera, Texas, by, 26
  quoted, 26, 27
blue whale, 70
Bogert, Charles M., 157
Bone Cabin Quarry, Wyoming, 138
Bowler, J. Kevin, 156
Brachiosaurus, xiii, 36, 70, 98
  distribution of, 82
brain to body ratios, 95
Brazos River, 8
Broili, Ferdinand, 156
brontosaurs and turtles, 126
Brontosaurus
  attacked by female allosaur, 122–23
  basking of, 48–49
  brain and body weight of, 95
  builds nest and lays eggs, 108–13
  cooled by rain, 65
  description of, 3–4
  disturbs pantotheres, 52
  at edge of cliff, 69
  escapes from Allosaurus, 8
  escapes from flood, 133–34
  feeding by, 50–52, 54–55
  feeding in swamp and jungle by, 14, 15, 20
  footprints of, 10–11
  leads herd after allosaur attack, 92
  locomotion of, 67–68
  looks at landscape, 85–88
  mating of, 101
  mechanics of hearing in, 87–88
  meets stegosaurs, 124–25
  method of feeding of, 16–17
  at northern sea, 76–77
  pattern of footfalls of, 66–67
  poling and swimming of, 19
  possible color vision in, 86–87
  possible delay between mating and nesting of, 114
  protected by size, 33–34
  rests in swamp, 13
  returns to nesting site, 141–45
  scent of, 88
  temperature adjustments in water of, 51–52, 54–55
  tracks of, 5
  unconcern for hatchlings of, 150–51
  use of neck of, 21
  wandering of, 80
Brontosaurus, synonymous with *Apatosaurus*, 154

Camarasaurus
  differences from Brontosaurus, 34–35
  distribution of, 82
  juvenile skeleton of, 148
  teeth of, 22, 35
Camptosaurus, 38–39
carnivorous dinosaurs, method of feeding of, 16
Carr, Archie, 103
Ceratosaurus, 38
Colbert, Edwin H., 154, 157, 161–62
color vision in reptiles, 93
Conant, Roger, 155, 156
conifers, 1, 2
continental connections, 81–82
Coombs, Walter P., Jr., 162
Cott, Hugh B., 159, 160, 164
  on choice of nesting site by crocodiles, 115
  on courtship patterns in crocodiles, 104
  on egg clutches in crocodiles, 117

on female crocodile assisting hatchlings, 127
on guarding of nests by crocodiles, 125
on method of egg laying in crocodiles, 117
on seclusion of young crocodiles, 149–50
on size of crocodile nests, 115
Cowles, Raymond B., 157
Cretaceous Period, dates of, 154
crocodiles
attacks by, 43
female, carrying of young in mouth, 127
nests of, 116
vision of, 93–94
crocodilians
courtship patterns in, 104–5
distribution of, 57
ectothermic, 56
temperature tolerances of, 57–58
cycads, 2, 86

Dalzell, Bonnie, 157
dating rocks, 153
Desmond, Adrian J., 157
Dinosaur National Monument, 136, 138, 160
dinosaurs
alternative possibilities of body temperatures of, 62–63
eggs of, 116
herds of, 44–45
orders of, xii
possible endothermism in, 60–62
temperature controls of, 56–63
Diplodocus, 36–37
nostrils of, 23, 37, 98
teeth of, 22, 37
Douglas-Hamilton, Iain and Oria, 80, 157, 158, 164
Douglass, Earl, 137
duck-billed dinosaur, food consumption of, 25

eardrum in dinosaurs, 95
ectothermic reptiles, 56
ectothermic vertebrates, defined, 153
Edmund, Gordon, 162
elephants
aggressiveness of toward lions, 42
behavior of related to weight, 73
daily food intake of, 155–56
damage to forest by, 29, 80
ratio of daily food consumption to weight in, 24
endothermic reptiles, 56
endothermic vertebrates, defined, 153
endothermism in dinosaurs, 59–63
Enlow, Donald H., and Sidney O. Brown, 158
on bone histology in dinosaurs, 61–62
extrastapes in dinosaurs, 94

Feduccia, Alan, 157
fenestra ovalis, 95
ferns, 2

Galapagos tortoise, ratio of daily food consumption to weight in, 25
Gans, Carl, 159
giantism, advantages of, 41–44
Gilmore, C. W., 162–63
giraffe
hyperventilation in, 31
rete mirabile of, 31
Glen Rose, Texas, 8, 26, 80
Glen Rose limestone (or Formation), 8–11, 27–28
Gray, James, 158

hair on flying reptiles, 156
Hatcher, J. B., 163
hatching of baby sauropods, 145–47
Haversian system in dinosaurs, 62
Heptonstall, W. B., 159
on gliding of *Archaeopteryx*, 128
heterothermic reptiles, 56
histology of dinosaur bone, as clue to body temperatures, 61–62

horsetails, 2, 86
Hotton, Nicholas, III, 162
hyposphene-hypantrum, 71–72
Hypselosaurus eggs, 116, 148

Jurassic Period, dates of, 154
juvenile sauropods join herd, 142–44

Keen, Paul, 156
Kentrosaurus, 82
Komodo lizard, 60

Langston, Wann, on Texas tracks, 154–55
Laws, R. M., 155
ligamentum nuchae, 21
lizards, 39

McIntosh, J. E., 163
mammalian ear bones, 95
marine turtle hatchlings race to sea, 149
mating of sauropods, 100–102
Megalosaurus, 82
Mesozoic Era, defined, xi–xii
migrations of modern mammals, 103
migrations of sea turtles, 102–3
Minton, Sherman A., and Madge Rutherford, 164
Moodie, Roy L., 159
Mook, C. C., 163
Morrison formation, xiii, 157

Neill, Wilfred T., 157, 159, 164
  on attacks by alligators, 43–44, 94
  on courtship patterns in crocodilians, 104, 105
  on guarding of nest by female alligators, 126
  on incubation of eggs in alligators, 114–15
New Zealand, 40

Omosaurus, 82
Opisthias, 40
Ornithischia, xii
Ornitholestes, 39, 50, 69, 98, 149
Osborn, H. F., 163
Ostrom, John H., 156, 157, 159, 163
  on differences of dinosaurs from other reptiles, 60
  on gregariousness of dinosaurs, 27
  on use of wings by *Archaeopteryx*, 128

Paluxy River, 8–9
pantotheres, 52–54
Parkinson, John, 160
  on deposition of dinosaurs at Tendaguru, 139
pinna, 95
Pleurocoelus, as possible source of Texas tracks, 154–55
Pooley, Anthony C., 159
potholes, 9
predation on hatchling sauropods, 146–47, 149
Protoceratops, nest and eggs of, 116–17
pterosaurs, presence of fur in, 40

rete mirabile in giraffe, 31
rhynchocephalians, 39–40
Ricqlès, Armand de, 158
  on bone histology in dinosaurs, as clue to body temperatures, 61–62

Saurischia, composition of, xii
sauropod carcasses, 134–36
sauropod hatchlings race to forest, 149
sauropod herd
  arrives at southern sea, 107
  attacked by allosaurs, 89–92, 99
  basking of, 49
  caught by flood, 132–34
  crosses embayment, 18–19
  damage to forest by feeding by, 18
  at edge of cliff, 68–69
  enters river valley, 131–32
  evidence of, from track findings, 26–27
  feeding by, 14–15, 20, 50

flees from female allosaur, 122–23
matings in, 100–102
meets another herd, 78–79
meets female allosaur, 119
meets juvenile sauropods, 142
meets stegosaurs, 123–24
migration of, 102–3
nesting of, 108–13
returns to nesting site, 141
returns to south, 97–100
at shore of northern sea, 75–77
spends night in swamp, 54–55
sauropods
abundance of fossils of, 45
adaptations in skeletons to weight of, 70–73
at apogee, xiv
blood circulation in, 31
breathing in, 31
conservation of energy in, 24–26
debates as to poses in, 158
devastation of jungle by, 29, 80
function of tail of, 72
giantism in, 70–73
limbs of, 28
locomotor analogies with elephants and, 29
long neck of, 30–31
neck of, as deterrent to attack, 44
possible courtship patterns in, 103–5
teeth of, 22
theories concerning feeding of, 23
variety of, 34–37
wide distribution of, 81–83
Schaeffer, Bobb, 158
Sharov, A. G., 156
Solnhofen limestone, 156–57
Sordes pilosus, 156
sperm storage in reptiles, 114
Sphenodon, 40
Špinar, Zdeněk V., 162
stapes in dinosaurs, 94
stegosaurs
brain and behavior of, 127
confrontation of, with brontosaurs, 124
Stegosaurus, 37, 39, 82
size of brain of, 44
stomach stones, 17
Sundance Sea, xiii, 76–77, 80–81
Swinton, W. E., 162

Tendaguru, Tanzania, 139
tree-ferns, 2, 86
Triassic Period, dates of, 154
Tsavo National Park, Kenya, 80

Warren, James V., 156
Wassersug, Richard J., 157
White, Theodore E., 159, 162
Williamsonias, 2, 52
Wilson, Woodrow, 137